PIEZOELECTRIC AEROELASTIC ENERGY HARVESTING

PIEZOELECTRIC AEROELASTIC ENERGY HARVESTING

HASSAN ELAHI
MARCO EUGENI
PAOLO GAUDENZI
Department of Mechanical and Aerospace Engineering
Sapienza University of Rome
Rome, Italy

Elsevier
Radarweg 29, PO Box 211, 1000 AE Amsterdam, Netherlands
The Boulevard, Langford Lane, Kidlington, Oxford OX5 1GB, United Kingdom
50 Hampshire Street, 5th Floor, Cambridge, MA 02139, United States

Copyright © 2022 Elsevier Inc. All rights reserved.

MATLAB® is a trademark of The MathWorks, Inc. and is used with permission.
The MathWorks does not warrant the accuracy of the text or exercises in this book.
This book's use or discussion of MATLAB® software or related products does not constitute endorsement or sponsorship by The MathWorks of a particular pedagogical approach or particular use of the MATLAB® software.

No part of this publication may be reproduced or transmitted in any form or by any means, electronic or mechanical, including photocopying, recording, or any information storage and retrieval system, without permission in writing from the publisher. Details on how to seek permission, further information about the Publisher's permissions policies and our arrangements with organizations such as the Copyright Clearance Center and the Copyright Licensing Agency, can be found at our website: www.elsevier.com/permissions.

This book and the individual contributions contained in it are protected under copyright by the Publisher (other than as may be noted herein).

Notices

Knowledge and best practice in this field are constantly changing. As new research and experience broaden our understanding, changes in research methods, professional practices, or medical treatment may become necessary.

Practitioners and researchers must always rely on their own experience and knowledge in evaluating and using any information, methods, compounds, or experiments described herein. In using such information or methods they should be mindful of their own safety and the safety of others, including parties for whom they have a professional responsibility.

To the fullest extent of the law, neither the Publisher nor the authors, contributors, or editors, assume any liability for any injury and/or damage to persons or property as a matter of products liability, negligence or otherwise, or from any use or operation of any methods, products, instructions, or ideas contained in the material herein.

Library of Congress Cataloging-in-Publication Data
A catalog record for this book is available from the Library of Congress

British Library Cataloguing-in-Publication Data
A catalogue record for this book is available from the British Library

ISBN: 978-0-12-823968-1

For information on all Elsevier publications
visit our website at https://www.elsevier.com/books-and-journals

Publisher: Matthew Deans
Acquisitions Editor: Brian Guerin
Editorial Project Manager: Fernanda A. Oliveira
Production Project Manager: Kiruthika Govindaraju
Designer: Matthew Limbert

Typeset by VTeX

Ride the wild wind (Hey, hey, hey, hey)
Ride the wild wind (Hey, hey)
Gonna ride the wild wind
It ain't dangerous, enough for me

(Ride the wild wind – Queen)

Contents

List of figures	*xiii*
Biography	*xix*
Preface	*xxiii*
Acknowledgment	*xxv*

Part 1. Introduction

1. Piezoelectric material — **3**
- 1.1. Introduction — 3
- 1.2. The piezoelectric effect — 5
  - 1.2.1. Direct effect of piezoelectricity — 5
  - 1.2.2. Converse effect of piezoelectricity — 7
  - 1.2.3. Constitutive equations of piezoelectricity — 8
  - 1.2.4. Open- and closed-circuit condition under compression — 9
  - 1.2.5. Energy coupling coefficients — 10
- 1.3. Piezoelectric materials for energy harvesting — 11
  - 1.3.1. Need of piezoelectric materials — 11
  - 1.3.2. FEA for piezoelectric materials — 12
  - 1.3.3. Energy generation — 13
  - 1.3.4. Energy utilization — 15
- References — 15

2. Smart structures — **21**
- 2.1. Introduction — 21
- 2.2. Piezoelectric smart structures — 22
- 2.3. Shape memory alloys — 22
  - 2.3.1. Electrorheological (ER) fluids — 23
  - 2.3.2. Magnetorheological (MR) fluids — 23
- 2.4. Piezoelectric sensors — 24
- 2.5. Sandwich structures — 26
- 2.6. Piezothermoelastic materials — 27
- 2.7. Piezothermoelastic response governing equations — 28
- 2.8. An example of smart structures application: machine learning based damage assessment — 30
  - 2.8.1. Laminated composites damage assessment with machine learning — 31
- References — 34

vii

viii   Contents

Part 2.   Energy harvesting

3.   Energy harvesting — 41
3.1.   Introduction — 41
3.2.   Sources for energy harvesting — 43
    3.2.1.   Piezoelectric — 44
    3.2.2.   Kinetic — 44
    3.2.3.   Photovoltaics — 46
    3.2.4.   Thermoelectrics — 46
    3.2.5.   Antennas — 46
    3.2.6.   Vibrations — 46
    3.2.7.   Wind — 46
3.3.   Mechanical energy harvesting — 46
    3.3.1.   Mechanical modulation principle — 47
3.4.   Fluid–structure interaction — 48
3.5.   Thermal energy — 49
3.6.   Photovoltaic technology — 50
3.7.   Acoustic energy — 51
3.8.   Radio frequency energy — 53
3.9.   Security threats for energy harvesting system — 54
References — 55

4.   Piezoelectric energy harvesters — 61
4.1.   Introduction — 61
4.2.   Piezoceramics-based energy harvesting — 62
    4.2.1.   Cymbal type — 64
    4.2.2.   Cantilever-type vibration energy harvesting — 65
    4.2.3.   Modeling and theory — 65
    4.2.4.   New materials for energy harvesting — 67
4.3.   Energy harvesting with piezopolymers — 68
4.4.   Harvesting model — 70
    4.4.1.   Governing equations — 70
References — 74

5.   Energy harvesting and circuits — 79
5.1.   Introduction — 79
5.2.   Piezoelectric energy harvesting circuits — 81
5.3.   Energy conditioning circuits — 82
    5.3.1.   Synchronous circuits — 82
    5.3.2.   Circuitry for enhanced energy harvesting — 84
5.4.   Equivalent circuit method — 85
    5.4.1.   Design of piezoelectric systems — 86
    5.4.2.   Power calculation for piezoelectric energy harvesting — 86

Contents ix

5.5.	Impedance method circuit		87
	5.5.1.	MPPT for piezoelectricity	88
	5.5.2.	SSHI and SECE	91
	5.5.3.	Rectifiers	91
	5.5.4.	Voltage doublers	93
References			93

6. Modeling and simulation of a piezoelectric energy harvester — 99

6.1.	Introduction	99
6.2.	Modeling	100
6.3.	Material creation	103
6.4.	Material assignment	104
6.5.	Interaction	107
6.6.	Input step creation	108
6.7.	Output assignment	110
6.8.	Boundary conditions	112
6.9.	Loading conditions	112
6.10.	Meshing	113
6.11.	Job creation	117
6.12.	Results	117
References		120

Part 3. Aeroelastic energy harvesting

7. Fluid–structure interaction: some issues about the aeroelastic problem — 125

7.1.	Introduction		125
	7.1.1.	Basic definitions of stability	126
7.2.	Bifurcation problems		127
	7.2.1.	Hopf bifurcation	128
7.3.	Aeroelastic problem formulation		128
	7.3.1.	$p–k$ method	128
7.4.	Finite element method for flag-flutter		130
	7.4.1.	Numerical model	130
7.5.	2D modeling for steady FSI system		131
	7.5.1.	2D structures	131
	7.5.2.	2D aerodynamic surfaces	132
	7.5.3.	FSI system	133
7.6.	Dynamic FSI		133
7.7.	Aerodynamic theories		136
	7.7.1.	Wagner function	137
	7.7.2.	Kussner function	138
7.8.	General approximation		139

x Contents

7.8.1.	Strip theory	139
7.8.2.	Quasi-steady	139
7.8.3.	Slender body/wing	140
References		141

8. Flutter-based aeroelastic energy harvesting — 143

8.1.	Introduction	143
8.2.	Flutter analysis: classical approach	144
8.3.	Flutter solutions	146
8.4.	Aeroelastic energy harvesters based on flutters	147
References		153

9. Limit cycle oscillations — 157

9.1.	Introduction	157
9.2.	Non-linear aeroelastic system	160
9.3.	Non-linear aeroelastic systems's LCO	162
9.3.1.	Numerical modeling of LCOs	163
9.3.2.	Non-linear aeroelastic model	164
9.4.	Theoretical modeling of aeroelastic harvester	168
9.5.	Aerodynamic modeling of aeroelastic harvester	169
9.6.	Structural model	172
9.6.1.	Non-linear electroelastic equation of motion	173
9.6.2.	Aeroelectroelastic state space equations	175
References		176

10. Vortex-induced vibrations based aeroelastic energy harvesting — 181

10.1.	Introduction	181
10.2.	Simple numerical example for a VIV aeroelastic energy harvester	183
10.2.1.	Existence of a "critical mass"	184
10.3.	Vortex-induced vibrations in circular cylinders	185
10.4.	Energy harvesting form the PZT based on VIV	185
10.5.	VIV-based energy harvesters	189
10.5.1.	Circular cylinder based VIV energy harvesters	190
10.5.2.	Energy harvester with turbulent flow	190
10.5.3.	Usage of poled and electrode flexible ceramic cylinder for power harvesting	191
10.5.4.	Energy harvesting from a rigid, elastically mounted spherical cylinder	191
10.5.5.	Energy harvester composed of aluminum cantilevered shim	191
10.5.6.	Energy harvester based on electromagnetic induction	192
10.5.7.	Vortex-induced vibration aquatic clean energy	192
10.5.8.	Passive turbulence control	192
10.5.9.	Harvester's performance enhancement	192

Contents **xi**

10.5.10. Harvesting energy from rigid circular cylinder 194
References 195

11. Galloping-based aeroelastic energy harvesting 201
11.1. Introduction 201
11.2. Transverse galloping 202
    11.2.1. Quasi-steady estimation 203
11.3. Mathematical model of transverse galloping 208
11.4. Wake galloping 209
    11.4.1. Types of wake galloping 211
    11.4.2. Turbulence effects 211
    11.4.3. Galloping response 212
11.5. Conversion factor 212
11.6. Evaluation of critical conditions with refined and multi-model approaches 213
11.7. Semi-analytical versus numerical solutions: non-linear galloping 215
11.8. Harnessable energy 216
References 216

12. Experimental aeroelastic energy harvesting 223
12.1. Introduction 223
    12.1.1. Ground vibrational tests 224
12.2. Experiments in a wind tunnel 225
    12.2.1. Testing of sub-critical flutter 226
    12.2.2. Flutter boundary 228
    12.2.3. Safety devices 231
    12.2.4. Research testing versus clearance testing 232
    12.2.5. Laws for scaling 232
    12.2.6. Flight experiments 233
    12.2.7. Flutter boundary approach 234
12.3. Testing of the flight flutter 234
    12.3.1. Excitation 234
12.4. Role of theory and experimentation in design 236
12.5. Experimental wing model 237
    12.5.1. Structural model 239
12.6. Slender body theory 240
12.7. Correlation between theory and experiment 241
12.8. Volterra theory experimentation 242
References 243

13. Concluding remarks 247

Subject index 249

List of figures

Figure 1.1	Piezoelectric effect in quartz.	4
Figure 1.2	Cubic structure of $BaTiO_3$: the red spheres represent the oxide centers, blue spheres represent the Ti^{4+} cations, and the green spheres represent Ba^{2+}.	4
Figure 1.3	Overview of PEH mechanism: (a) piezoelectric effect with 33 and 31 strain–charge couplings; (b) polarization process; and (c) bimorph PEH mechanism.	5
Figure 1.4	Electromechanical coupling in piezoelectric effect.	5
Figure 1.5	Piezoelectric materials for sensing and actuating: (a) S–E field plot, (b) hysteresis plot of P–E, (c) piezoelectric materials before and after poling, (d) applied voltage and polarity voltage directions are in the same direction, (e) applied voltage and polarity voltage directions are in different directions, (f) generated voltage and polarity voltage are in a different directions during compression, and (g) generated voltage and polarity voltage are in different directions during tension.	6
Figure 1.6	Undeformed and deformed configurations under the effect of σ_3.	6
Figure 1.7	Non-deformed and deformed configurations under the action of E_3.	7
Figure 1.8	Open- and closed-circuit configurations of a piezoelectric material under compression; D is electric flux and E is electric field.	9
Figure 1.9	The path of mechanical loading considering different circuit configurations.	10
Figure 1.10	The path of electrical loading considering different circuit configurations.	11
Figure 1.11	Typical cantilever piezoelectric harvesters; V is the voltage that can be harvest from this PEH.	13
Figure 1.12	Overall mechanism for piezoelectric energy harvesting; K is the modal stiffness, C is the modal damping, C_e is the external capacitance, and V is the output voltage generated.	14
Figure 1.13	Application of PEH in wireless sensors.	14
Figure 2.1	Machine learning, its objectives, and commonly used algorithms.	30

xiv List of figures

Figure 2.2	A concise explanation of the damage classification procedure [80].	32
Figure 3.1	Energy sources.	43
Figure 3.2	Renewable sources of energy.	43
Figure 3.3	Renewable sources of energy harvesting in aerospace industry.	44
Figure 3.4	Mechanisms of energy harvesting: light, temperature, motion, or electromagnetic (EM) field are exposed to energy processors which convert them into useful electrical energy.	44
Figure 3.5	Various sources of energy harvesting.	45
Figure 3.6	Various sources of mechanical energy harvesting.	47
Figure 3.7	Various mechanisms for fluid–structure interaction.	49
Figure 3.8	Solar cell operation.	51
Figure 3.9	Schematic diagram for overall acoustic energy harvesting.	52
Figure 3.10	Sources for RF energy harvesting.	54
Figure 3.11	Security threats for energy harvesting system.	55
Figure 4.1	Conventional PZT energy harvesters.	64
Figure 4.2	Conventional PZT energy harvesters.	65
Figure 4.3	Geometry and position of the neutral axis of PCGE-A.	68
Figure 4.4	An equivalent model for a piezoelectric vibration energy harvesting system.	71
Figure 4.5	A typical AC–DC harvesting circuit.	71
Figure 5.1	Energy harvesting circuit that utilizes energy on the spot.	81
Figure 5.2	Energy harvesting circuit that stores energy.	81
Figure 5.3	Aeroelastic piezoelectric energy harvesting circuits [37].	84
Figure 5.4	Equivalent circuit method for piezoelectric energy harvester.	86
Figure 5.5	Hill climbing algorithm for energy harvesting.	88
Figure 5.6	FOCV algorithm for energy harvesting.	90
Figure 5.7	SSHI circuit for a piezoelectric energy harvester.	91
Figure 5.8	Rectification circuit for a piezoelectric energy harvester.	92
Figure 5.9	Schematics of the circuit for voltage doublers.	92
Figure 6.1	Common steps carried out during modeling and simulation of a piezoelectric energy harvester.	100
Figure 6.2	Abaqus software home screen.	101
Figure 6.3	Model creation input parameters.	101
Figure 6.4	Grid size for model creation.	102
Figure 6.5	Geometry of the piezoelectric patch.	102
Figure 6.6	Extrusion of the model.	102
Figure 6.7	3D model of an energy harvester.	103
Figure 6.8	Material creation.	103
Figure 6.9	Assigning density to piezoelectric material.	104

Figure 6.10	Assigning elastic properties.	104
Figure 6.11	Orthotropic properties assignment.	105
Figure 6.12	Assigning strain values.	105
Figure 6.13	Anisotropic values.	105
Figure 6.14	Assigning stress values.	106
Figure 6.15	Create section.	106
Figure 6.16	Material assignment.	106
Figure 6.17	Material assigned.	107
Figure 6.18	Editing material orientation.	107
Figure 6.19	Material orientation rotation.	108
Figure 6.20	Assembly.	108
Figure 6.21	Create instances.	109
Figure 6.22	Input step for loading condition.	109
Figure 6.23	Time period of the input step.	109
Figure 6.24	Increment size for the input step time period.	110
Figure 6.25	Different options for step input.	110
Figure 6.26	Field output manager.	111
Figure 6.27	Electrical potential output.	111
Figure 6.28	Output creation for input step.	111
Figure 6.29	Boundary conditions creation.	112
Figure 6.30	Electric potential boundary condition.	112
Figure 6.31	Overall boundary conditions of the model.	113
Figure 6.32	Load creation for a piezoelectric energy harvester.	113
Figure 6.33	Loading parameters.	114
Figure 6.34	Overall load representation.	114
Figure 6.35	Global seeds for meshing.	116
Figure 6.36	Local seeds for meshing.	116
Figure 6.37	Overall meshing of the piezoelectric energy harvester.	116
Figure 6.38	Element type for meshing.	117
Figure 6.39	Job creation.	118
Figure 6.40	Stresses calculation of the piezoelectric energy harvester.	118
Figure 6.41	Electric potential generated by the piezoelectric energy harvester.	118
Figure 6.42	XY plot creation.	119
Figure 6.43	Electric potential vs time in graphical format.	119
Figure 6.44	Electric potential vs time in tabular format.	119
Figure 7.1	Collar's triangle for aeroelasticity.	126
Figure 7.2	Wagner function.	137
Figure 7.3	Kussner function.	138
Figure 8.1	Overall mechanism of piezoelectric energy harvesters [26].	148

xvi List of figures

Figure 8.2	Bifurcation diagram of the Al patched harvester [26].	149
Figure 8.3	Bifurcation diagram of the aeroelastic piezoelectric energy harvester [26].	149
Figure 8.4	Response to flutter velocity and frequency by the harvester of various length.	150
Figure 8.5	Output power of aeroelastic piezoelectric energy harvester; correlation factor of 0.92 between numerical and experimental results.	153
Figure 9.1	LCOs of an aeroelastic piezoelectric energy harvester via LCOs [19]; flutter occurs at 25 ms^{-1}.	163
Figure 9.2	LCOs of an aeroelastic harvester attached with Al patch via LCOs [19]; flutter occurs at 23 ms^{-1}.	164
Figure 9.3	Numerical deformation comparison for Al and PZT patched flags, from 15 to 38 cm long [19].	165
Figure 9.4	LCOs of the wing [20].	167
Figure 9.5	LCOs output variables: (a) amplitude of the pitch angle at the $\cap\theta_{\mathrm{tip}}$, (b) amplitude of the voltage generated of one harvester $\cap v_{ei}$, (c) mean power generated P_m [20].	168
Figure 9.6	Aeroelastic structure model with pitch and plunge degrees of freedom.	169
Figure 10.1	Schematic of a wake VIV-based piezoelectric energy harvester.	185
Figure 11.1	Overall mechanism of a galloping-based piezoelectric energy harvester [9].	202
Figure 11.2	Schematic of a galloping-based piezoelectric energy harvester.	203
Figure 11.3	Fluid force on the cross-section.	208
Figure 11.4	Schematic of a wake galloping-based piezoelectric energy harvester.	210
Figure 11.5	Output power of a galloping-based piezoelectric energy harvester [9].	216
Figure 12.1	Overall energy harvesting model of flag-flutter subjected to axial flow in sub-sonic wind tunnel [63].	226
Figure 12.2	Overall experimental campaign for flag-flutter energy harvesting mechanism carried out in sub-sonic wind tunnel [63].	227
Figure 12.3	Al and PZT attached flags used for energy harvesting in sub-sonic wind tunnel [63].	228
Figure 12.4	Deformation comparison for Al and PZT patched flags from 15 to 38 cm long [63].	229
Figure 12.5	Experimental output voltage for different resistances [63].	230
Figure 12.6	Flutter for reduced critical velocity [63], where $\mu = \rho_F L/\rho_S t$ and $U_{RC}/\mu = \left[(\rho_P h)^{3/2}/(\rho_F D^{1/2})\right] U$.	231

Figure 12.7	Flutter for reduced critical frequency [63], where $\mu = \rho_F L/\rho_S t$ and $F_C/\mu^2 = \left[(\rho_P h)^{3/2}/(\rho_F D^{1/2})\right]f_c$.	232
Figure 12.8	Experimental setup for aeroelastic energy harvesting via galloping mechanism [65].	233
Figure 12.9	Experimental campaign for VIV based piezoelectric energy harvesting [66].	234
Figure 12.10	Aircraft's flutter clearance process.	236
Figure 12.11	Physical representation of wing model.	238

Biography

Hassan Elahi

Dr. Hassan Elahi currently works as an Associate Researcher at the Department of Mechanical and Aerospace Engineering, Sapienza University of Rome, Italy. Previously he has worked as a Lecturer at the Department of Mechanical Engineering, Institute of Space Technology, Islamabad, Pakistan.

Dr. Elahi has completed his Ph.D. in the field of Nonlinear Piezoelectric Aeroelastic Energy Harvesting. He has received an Award of Honors in his Master's Degree program. Moreover, his name is included in the Foreign Minister Honor List 2021 by the Government of Pakistan for his services in Science and Technology. Furthermore, his name is included in 24 leaders of Pakistan 2021 under the age of 40 years by the Ministry of Foreign Affairs, Government of Pakistan.

Dr. Elahi research interests are in the field of smart structures, piezoelectric materials, aerospace structures, aeroelasticity, energy harvesting, and MEMS/NEMS.

Marco Eugeni

Dr. Marco Eugeni is an Associate Researcher at the Mechanical and Aerospace Engineering Department of the Sapienza University of Rome

where he teaches also the Structural Calculus Laboratory for the first level degree in Aerospace Engineering. His research activities are focused on aerospace structures, space systems, concurrent engineering and multidisciplinary methods for engineering design, multiphysics analysis, aeroelasticity, nonlinear dynamical systems, reduced-order modeling, and viscoelastic materials structural modeling. Recently his research activities are focusing on Smart Manufacturing for Space Industry and Additive Manufacturing.

Paolo Gaudenzi

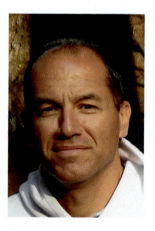

Prof. Paolo Gaudenzi is Full Professor of Aerospace Structures at Sapienza University of Rome and Director of the Mechanical and Aerospace Engineering Department and author of more than 150 papers and of the research book Smart structures, J. Wiley 2009. His research activities are focused on aerospace structures, space systems, concurrent engineering and multidisciplinary methods for engineering design, cost engineering, laminated and composite structures, active materials and intelligent structures, thermal structures, finite element modeling, multiphysics analysis. During the last years, his research is focusing also on Additive Manufacturing and applications of Smart Manufacturing to the Manufacturing Assembly Integration and Testing processes of space systems. Prof. Paolo Gaudenzi is also active in technological transfer from Academic to Industry being one of the founders of the start-up company Smart Structures Solutions s.r.l. He is also holder of 2 patents covering innovative technical solutions for the monitoring of integrity of infrastructures by using smart structures and smart composite concepts.

Preface

For the last few decades, piezoelectric materials have been playing a fundamental role in the field of energy harvesting. These materials are capable of driving micro-electromechanical systems by absorbing the energy from the surroundings thanks to their properties to generate a voltage differential once exposed to a deformation or stress field.

The increasing interest in the industrial world for a very large and distributed network of sensors for monitoring and IoT representations of systems and processes have brought energy harvesting in a central role for those applications where the energy supply can be a critical issue because of operative conditions, system complexity and a strict limit on the weight that harnessing and batteries can add to the specific system.

Thus, the strict requirements in the aerospace industry lead to the formulation of various novel energy harvesting mechanisms, one of which is the piezoelectric energy harvesting mechanism.

Piezoelectric transducers are used to describe the phenomenon of energy generation as the energy transformation from the operating environment into electric energy that can be used on the premises for actuation or deposited in batteries for use in the future. Due to lightweight engineering designs, leading to micro-/nano-powered electronic circuits, the world is changing to low-powered electronic devices. Many researchers concentrated on the self-powered use of the piezoelectric energy harvester over battery use. These harvesters are suitable for use in microelectronic systems, smart buildings, tracking of structural health, and as wireless sensor systems for sub-orbital missions during recent technical progress.

Many scientists have highlighted the relevance of piezoelectric device modeling. In linear and non-linear situations, for 3D solids, as well as structural elements, such as plate and shell, numerous technologies have been developed. By taking energy from the surroundings, they can harvest valuable electricity that can be used for electronics or deposited in batteries for later purposes. A piezoelectric aeroelastic energy harvester consumes airflow energy and transforms it into usable electrical energy, which is analyzed in this book.

When the Fluid–Structure Interaction (FSI) problem is considered it is necessary to consider the entire dynamics of the structure and flow together leading to the concept of the aeroelastic system later introduced in this

xxiii

book. From a mathematical point of view, this relation happens because the state of the structural system is dictated by the flow pressure whereas the fluid state is influenced by the structural system which is seen as a boundary condition for the flow governing equations.

This results in an inherently non-stationary and rather dynamic phenomenon and cannot be studied anymore by separate consideration of the structure and flow. It is self-evident that the mean velocity of the flow plays a critical role in the dynamics of an aeroelastic system where different kind of structural excitation and instability phenomena can happen. Both the flow excitation of the structures and the dynamical aeroelastic instability phenomena can be used for energy harvesting purposes by means of PZT devices and this will be deeply analyzed in the present book.

Thus, in this book, different strategies to harvest useful electrical energy by absorbing the energy from the surroundings via PZT materials are discussed in detail with a particular focus on those where the energy is gathered from a fluid–structure interaction. Particular emphasis is placed on demonstrating correct modeling for the unsteadiness of aerodynamics. Indeed, the aerodynamic model is a critical ingredient for a sound prediction of the linear behavior of an aeroelastic system: without correct aerodynamic modeling is not possible a correct evaluation of the nonlinear behavior which is at the base of most of the energy harvesters that will be analyzed.

These harvesters can be deployed in many locations, such as urban areas, high wind areas, ventilation outlets, rivers, ducts of buildings, lifting components in aircraft structures, etc. These harvesters can be used to power small electronic devices including health monitoring sensors, medical implants, data transmitters, wireless sensors, and cameras.

Hassan Elahi, Marco Eugeni, and Paolo Gaudenzi

Acknowledgment

The authors would like to thank and mention with sincere affection the colleagues that make possible the development of some of the original results presented in this book concerning the energy harvesting devices that use the flag flutter phenomenon. In particular, we would like to thank Prof. Franco Mastroddi for his advices in aeroelastic modeling, Prof. Luca Lampani for his knowledge of composite manufacturing and embedded electronics, and Prof. Giovanni Paolo Romano for make possible the experimentation in is laboratory and the advices for the experimental set-up. Moreover, the authors want to acknowledge the efforts of Dr. Aqsa Shakeel for her continuous support.

PART 1

Introduction

CHAPTER 1

Piezoelectric material

Contents

1.1.	Introduction	3
1.2.	The piezoelectric effect	5
	1.2.1 Direct effect of piezoelectricity	5
	1.2.2 Converse effect of piezoelectricity	7
	1.2.3 Constitutive equations of piezoelectricity	8
	1.2.4 Open- and closed-circuit condition under compression	9
	1.2.5 Energy coupling coefficients	10
1.3.	Piezoelectric materials for energy harvesting	11
	1.3.1 Need of piezoelectric materials	11
	1.3.2 FEA for piezoelectric materials	12
	1.3.3 Energy generation	13
	1.3.4 Energy utilization	15
References		15

1.1 Introduction

In this chapter the physics of the piezoelectric materials is elaborated. There exist certain materials which have the ability to generate electrical voltage by application of mechanical stress on them (i.e., the direct effect of piezoelectricity [1]), and vice versa (i.e., the converse effect of piezoelectricity). Pierre and Jacques Curie discovered the direct piezoelectric effect, whereas Gabriel Lippmann discovered the converse effect of piezoelectricity which was then experimentally verified by "the Curie brothers." The piezoelectric materials can generate electric voltage which is proportional to the input stress, and vice versa [1–4]. The piezoelectric materials belong to the crystalline group [5–7]. The topic of energy harvesting has become of significant importance over the last few decades, and the development of this field has been responsible for revolutionizing micro-electromechanical systems, i.e., wireless sensors [8]. The piezoelectric sensors and transducers that absorb ambient energy from vibrations are very much in demand because of their energy harvesting capability [9]. Many researchers are working on numerical simulations of piezoelectric-based sensors, actuators, and structural health monitoring systems based on the finite element method (FEM) [10]. The piezoelectric effect is considered to be a cou-

Piezoelectric Aeroelastic Energy Harvesting
https://doi.org/10.1016/B978-0-12-823968-1.00010-6

Copyright © 2022 Elsevier Inc.
All rights reserved.

4 Piezoelectric Aeroelastic Energy Harvesting

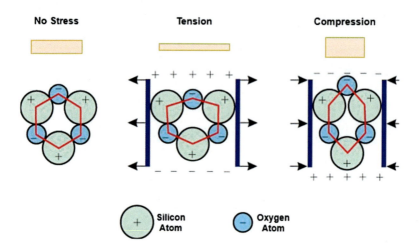

Figure 1.1 Piezoelectric effect in quartz.

pling between electrical and mechanical properties of a material [11]. The overall effect of piezoelectricity in quartz is represented in Fig. 1.1. A typical 3D crystal structure of barium titanate (BaTiO$_3$) ceramic is shown in Fig. 1.2. For enhancing the output electric potential, it is important to have a poling effect on the piezoelectric materials so that all the poles are in approximately the same direction. The overall mechanism of a bimorph PEH is presented in Fig. 1.3 with the elaboration of the polarization process.

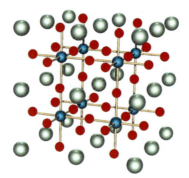

Figure 1.2 Cubic structure of BaTiO$_3$: the red spheres represent the oxide centers, blue spheres represent the Ti^{4+} cations, and the green spheres represent Ba^{2+}.

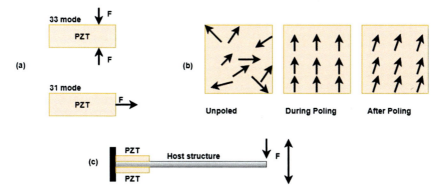

Figure 1.3 Overview of PEH mechanism: (a) piezoelectric effect with 33 and 31 strain–charge couplings; (b) polarization process; and (c) bimorph PEH mechanism.

1.2 The piezoelectric effect

The piezoelectric effect is defined as the generation of the non-mechanical output (electric potential) in response to the mechanical stimulus or mechanical output (mechanical strain). The collar's triangle for piezoelectricity is presented in Fig. 1.4. The effect of poling is elaborated in Fig. 1.5 for piezoelectric materials in sensors and actuators. Moreover, the hysteresis plot of polarization vs. electrical field and plot of strain vs. electric field are also shown in Fig. 1.5.

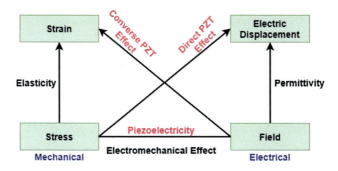

Figure 1.4 Electromechanical coupling in piezoelectric effect.

1.2.1 Direct effect of piezoelectricity

Piezoelectricity is defined as the mechanical stress along the polarization direction to generate an electric voltage. By the application of this mechan-

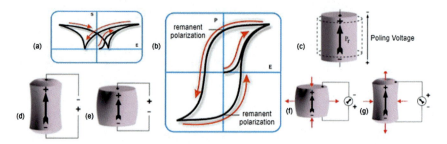

Figure 1.5 Piezoelectric materials for sensing and actuating: (a) S–E field plot, (b) hysteresis plot of P–E, (c) piezoelectric materials before and after poling, (d) applied voltage and polarity voltage directions are in the same direction, (e) applied voltage and polarity voltage directions are in different directions, (f) generated voltage and polarity voltage are in a different directions during compression, and (g) generated voltage and polarity voltage are in different directions during tension.

ical stress σ_3 along the x_3 axis, the material undergoes a cell deformation, an elongation along the stress application axis, as represented in Fig. 1.6.

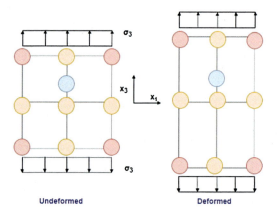

Figure 1.6 Undeformed and deformed configurations under the effect of σ_3.

The titanium atom, by changing its position, increases polarization which becomes $P_r + \Delta P$. For small values of σ_3,

$$\Delta P_3 = d_{33}\sigma_3, \tag{1.1}$$

where d_{33}[1] is a positive constant. Similarly, it is possible to define all the other constants. The direct effect of piezoelectric materials has vast appli-

---

[1] This notation links the polarization direction to the applied stress.

cations in the field of energy harvesting [12–15]. Considering all the applied stresses, the direct piezoelectric coupling matrix $\underline{\underline{d}}$ is defined as

$$\begin{Bmatrix} \Delta P_1 \\ \Delta P_2 \\ \Delta P_3 \end{Bmatrix} = \begin{bmatrix} 0 & 0 & 0 & 0 & d_{15} & 0 \\ 0 & 0 & 0 & d_{15} & 0 & 0 \\ d_{31} & d_{31} & d_{33} & 0 & 0 & 0 \end{bmatrix} \begin{Bmatrix} \sigma_1 \\ \sigma_2 \\ \sigma_3 \\ \sigma_4 \\ \sigma_5 \\ \sigma_6 \end{Bmatrix}. \qquad (1.2)$$

Eq. (1.2) can be written as

$$\Delta P = d\sigma. \qquad (1.3)$$

1.2.2 Converse effect of piezoelectricity

Piezoelectricity is defined as the application of an electric field along the polarization axis, which determines the displacement of positive charges in the same direction for a positive electric field applied. There is an elongation along the x_3 axis and a compression along x_1 and x_2 axes. Non-deformed and deformed configurations under the action of E_3 is represented in Fig. 1.7.

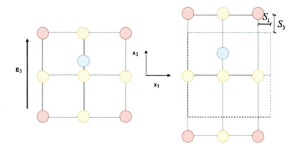

Figure 1.7 Non-deformed and deformed configurations under the action of E_3.

For the converse effect of piezoelectricity,

$$\begin{Bmatrix} \varepsilon_1 \\ \varepsilon_2 \\ \varepsilon_3 \\ \varepsilon_4 \\ \varepsilon_5 \\ \varepsilon_6 \end{Bmatrix} = \begin{bmatrix} 0 & 0 & d_{13} \\ 0 & 0 & d_{13} \\ 0 & 0 & d_{33} \\ 0 & d_{42} & 0 \\ d_{51} & 0 & 0 \end{bmatrix} \begin{Bmatrix} E_1 \\ E_2 \\ E_3 \end{Bmatrix}, \qquad (1.4)$$

8 Piezoelectric Aeroelastic Energy Harvesting

$$\varepsilon = dE. \tag{1.5}$$

1.2.3 Constitutive equations of piezoelectricity

By definition, the induced polarization in a dielectric material from the electric field vector is

$$P_i = \chi_{ij} E_j, \tag{1.6}$$

in which $\chi_{ij}(Fm^{-1})$ is the tensor of the dielectric susceptibility of the material.[2] The charge density induced in the material is obtained by introducing the electric displacement vector $D_i[Cm^{-2}]$, which is defined as

$$D_i = \epsilon_0 E_i + P_i, \tag{1.7}$$

where ϵ_0 represents vacuum dielectric permeability. Let us introduce the dielectric material permeability as

$$\epsilon_{ij} = \epsilon_0 \delta_{ij} + \chi_{ij}. \tag{1.8}$$

Substituting (1.6) and (1.8) into (1.7) gives

$$D_i = \epsilon_0 E_i + P_i = (\epsilon_0 \delta_{ij} + \chi_{ij}) E_j = \epsilon_{ij} E_j. \tag{1.9}$$

The presence of an additional polarization $\underline{\Delta P}$ due to the direct piezoelectric effect can be seen as an additional term of $\underline{\underline{D}}$ which can therefore be written as

$$\underline{D} = \underline{\underline{\epsilon}}^{(\sigma)} \underline{E} + \underline{\underline{d}}\,\underline{\sigma}, \tag{1.10}$$

where $\underline{\underline{\epsilon}}_\sigma$ is the dielectric matrix with constant or zero stress. It is also possible to write the deformation vector by adding a term due to the piezoelectric effect of the material:

$$\underline{\varepsilon} = \underline{\underline{Q}}^{(E)} \underline{\sigma} + \underline{\underline{d}}^T \underline{E}, \tag{1.11}$$

where the material compliance matrix is indicated by $\underline{\underline{Q}}$.

Unifying (1.10) and (1.11), the constitutive matrix of a continuous piezoelectric material is

$$\begin{Bmatrix} D \\ \epsilon \end{Bmatrix} = \begin{bmatrix} \varepsilon^{(\sigma)} & d \\ d^T & Q^{(E)} \end{bmatrix} \begin{Bmatrix} E \\ \sigma \end{Bmatrix}, \tag{1.12}$$

[2] Eq. (1.6) is valid only for linear materials or up to the linear limit for non-linear materials.

or in a tensorial way,

$$D_i = \varepsilon_{ij}^{\sigma} E_j + d_{ijk}\sigma_{jk},$$
$$\epsilon_{ij} = F_{ijhk}^{E}\sigma_{hk} + d_{ijk}^{t} E_i. \qquad (1.13)$$

It is important to introduce a coefficient that links the mechanical stress applied to the electric field produced [16–18]. This, together with d, allows characterizing the piezoelectric material applicability in order to take full advantage of its properties. Defining the coefficient g as

$$g_{ij} = d_{ij}\epsilon_{ij} \qquad (1.14)$$

yields

$$E_i = g_{ij}\sigma_j = \frac{d_{ij}}{\epsilon_{ij}}\sigma_j. \qquad (1.15)$$

Once the available electric field is established, the best use of the material in question can be identified according to d or g values. A high d is optimal for the actuation applications, whereas a high g maximizes sensor performance of the material.

1.2.4 Open- and closed-circuit condition under compression

For compression in piezoelectric materials, two types of configuration can be used for the phenomenon of energy harvesting, i.e., open- and closed-circuit configurations. In the open-circuit condition, the charges are disposed of and are not free to move towards electrodes, which results in no electric flux, whereas in the closed-circuit configuration, both electrodes have the same electric potential, resulting in no electric field. The phenomenon of open- and closed-circuit conditions of piezoelectric harvesters under compression is presented in Fig. 1.8.

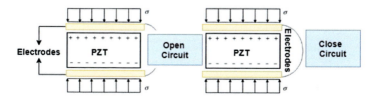

Figure 1.8 Open- and closed-circuit configurations of a piezoelectric material under compression; D is electric flux and E is electric field.

1.2.5 Energy coupling coefficients

It is important to evaluate the efficiency of the PEH. It can be calculated by evaluating the ratio of electrical work W_1 to the mechanical work W_2 with respect to electrical work, i.e., $W_1 + W_2$. The path of mechanical loading considering different circuit configurations is expressed in Fig. 1.9, and the path of electrical loading considering different circuit configurations is shown in Fig. 1.10. The coupling coefficient K_{33} for the open-circuit condition can be expressed as

$$K_{33} = \sqrt{\frac{W_1}{W_1 + W_2}}. \tag{1.16}$$

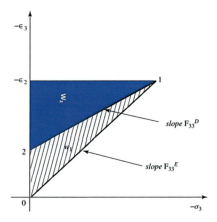

Figure 1.9 The path of mechanical loading considering different circuit configurations.

For mechanical loading, by taking into account Fig. 1.9, the energy coupling coefficient can be expressed as

$$K_{33} = \sqrt{\frac{F_{33}^E - F_{33}^D}{F_{33}^E}} = \sqrt{\frac{d_{33}^2}{F_{33}^E \epsilon_{33}^\sigma}}, \tag{1.17}$$

while K_{31}, the energy coupling coefficient, can be expressed as

$$K_{31} = \sqrt{\frac{d_{31}^2}{F_{11}^E \epsilon_{33}^\sigma}}. \tag{1.18}$$

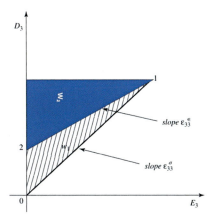

Figure 1.10 The path of electrical loading considering different circuit configurations.

1.3 Piezoelectric materials for energy harvesting

Energy harvesting via piezoelectric transduction can be a suitable alternative energy source rather than a traditional energy harvesting mechanism as it is applicable to power wireless sensors and actuators directly. This mechanism has numerous applications ranging from aerospace to civil engineering [19]. PEH is usually compact from a design point of view and has an advantage over traditional energy harvesting mechanisms by being environmentally friendly. Therefore, piezoelectric materials have applications in numerous biomedical devices [20], optical devices [21], mechanical/civil structures [22], space missions [23], and precise measurement tools [24].

Piezoelectric materials are suitable for energy harvesting as they are flexible with high torque-to-volume ratio [25,26]. The idea behind the piezoelectric energy harvesting is composed of three basic steps: (i) piezoelectric material absorbs the energy from surroundings, the source of energy can be structural vibrations, and can convert it into useful electrical energy; (ii) the storage unit on which the energy harvested can be stored, i.e., batteries; (iii) the external circuit or ICs which can convert this AC into DC [27]. Moreover, the piezoelectric materials can be used directly to drive electronic circuits; in this case, step (ii) can be skipped [27,28].

1.3.1 Need of piezoelectric materials

Design robustness of self-powered electronic devices has made their construction quite demanding these days [1]. As batteries are heavy and expensive to maintain, there is a need for a mechanism that can power the nano- or microelectronics by absorbing structural energy [19]. There are many mechanisms for energy transformation, i.e., electromagnetic, electromechanical, and fluid–structure interaction systems [12–15,29–35]. Among these mechanisms, electromechanical systems play a vital role because of their voltage-dependent actuation [36]. Many researchers are working on such techniques that drive electronic circuits in electromechanical systems [12,28]. The piezoelectric material is mostly used to harvest electrical energy by absorbing mechanical energy (mechanical vibrations) from the surroundings [6].

The importance of alternate energy sources arises as the lifespan of batteries is limited compared to the circuit life. In some cases, the maintenance/replacement of the battery is very costly or even impossible, e.g., for the systems that are being used in harsh conditions, i.e., cold or hot climate, high-altitude places, and icy or snowy regions, which can reduce battery life. In many cases, even battery recycling is a problem as it can cause environmental hazards. Therefore, an alternate energy harvesting mechanism is proposed, which can directly power the electronic devices and can even increase the battery life [19,37]. Moreover, in this chapter PEH is emphasized to be used as an alternate source of energy for aerospace applications.

1.3.2 FEA for piezoelectric materials

The finite element method for piezoelectric materials can be accessed in many commercially available software packages to perform static and dynamic analysis. In these numerical procedures, the structural elements of the material are coupled with the electrical properties of the piezoelectric material. FEA is a very attractive tool for modeling and simulation of electromagnetic and electromechanical sensors and acquisitions [16,38]. A.H. Allik introduced the first numerical simulation of a piezoelectric material in 1970 [39]. The output power of PEH plates was also predicted with an electromechanical coupled FE model. The goal was to power small electronic devices by transforming the waste vibration energy available in the atmosphere into electrical energy. The advantage of FEA over analytical results is that mechanical stress variations and electrical field calculations of complex geometries of the material can be more readily calculated. FEA was performed to calculate the stress and electric field distributions under static loads and under any electrical frequency. Thus the impact of material geometry can be evaluated and improved without the need to make and test various materials [40]. In addition, FEA can likely predict lifetime expectations without any need to conduct time-consuming tests if significant electrical and mechanical parameters are obtained.

1.3.3 Energy generation

The piezoelectric-based devices are lightweight and are capable of self-dependent actuation [41]. The efficiency of harvested energy for the direct effect of piezoelectricity can be analyzed by calculating the difference of mechanical energy converted in to electrical energy and loss in energy conversion [42–45]. This harvesting mechanism is dependent on the medium of interaction [46–49], i.e., transformation of kinetic energy to piezoelectric transducer [50–52]. This medium may be mechanical vibrations [53–55], fluid–structure interaction [56,57], and thermal interaction [58,59].

The cantilever beam is the most widely used configuration illustrating energy harvesting techniques in electromechanical systems. This configuration has wide applications in the field of sensors, actuators, structural health monitoring systems, and energy harvesters [23]. In this configuration, a piezoelectric material is attached to the beam, which is fixed on one end. If the number of piezoelectric patch is one then it is known as a unimorph cantilever harvester. The number of piezoelectric patch defines the configuration of piezoelectricity, i.e., bimorph or trimorph. When these harvesters are exposed to vibrations, mechanical energy is absorbed by piezoelectric patch and transformed into useful electrical energy [7]. The mechanism of a typical cantilever piezoelectric harvester is represented in Fig. 1.11.

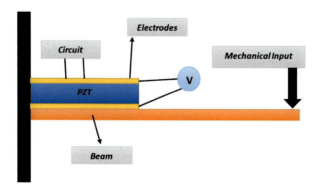

Figure 1.11 Typical cantilever piezoelectric harvesters; V is the voltage that can be harvest from this PEH.

14 Piezoelectric Aeroelastic Energy Harvesting

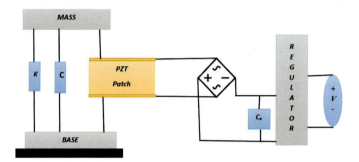

Figure 1.12 Overall mechanism for piezoelectric energy harvesting; K is the modal stiffness, C is the modal damping, C_e is the external capacitance, and V is the output voltage generated.

The piezoelectric cantilever harvesters follow the direct effect of piezoelectricity, i.e., mechanical input results in electrical output. When these harvesters are subjected to mechanical loading, i.e., vibrations, the charge is collected on the surface of piezoelectric patch causing the voltage difference between the thickness of the patch. There is a need to improve sensitivity or energy efficiency of the PEH mechanism in order to get higher voltage for same excitation [60]. The overall mechanism for piezoelectric energy harvesting is represented in Fig. 1.12.

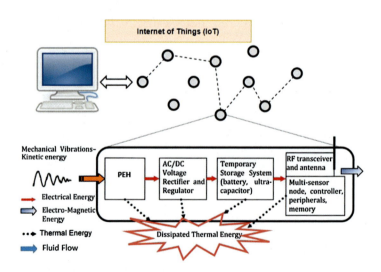

Figure 1.13 Application of PEH in wireless sensors.

1.3.4 Energy utilization

Energy harvested by a piezoelectric material can be used directly to operate sensors and actuators or can be stored in batteries for later purposes. These harvesters have many applications in the field of aerospace engineering, IoT, mechanical engineering, civil engineering, and even in wearables [23,36, 61,62]. Many researchers have been working on this technology for the last few decades. Now they have applications ranging from airplane structural health monitoring to the pacemaker (artificial heart). Moreover, one of the applications of PEH in wireless sensors is elaborated in Fig. 1.13.

References

[1] Paolo Gaudenzi, Smart Structures: Physical Behaviour, Mathematical Modelling and Applications, John Wiley & Sons, 2009.

[2] Mohsin Ali Marwat, Weigang Ma, Pengyuan Fan, Hassan Elahi, Chanatip Samart, Bo Nan, Hua Tan, David Salamon, Baohua Ye, Haibo Zhang, Ultrahigh energy density and thermal stability in sandwich-structured nanocomposites with dopamine@Ag@BaTiO$_3$, Energy Storage Materials 31 (2020) 492–504.

[3] Ahsan Ali, Riffat Asim Pasha, Hassan Elahi, Muhammad Abdullah Sheeraz, Saima Bibi, Zain Ul Hassan, Marco Eugeni, Paolo Gaudenzi, Investigation of deformation in bimorph piezoelectric actuator: analytical, numerical and experimental approach, Integrated Ferroelectrics 201 (1) (2019) 94–109.

[4] Muhammad Usman Khan, Zubair Butt, Hassan Elahi, Waqas Asghar, Zulkarnain Abbas, Muhammad Shoaib, M. Anser Bashir, Deflection of coupled elasticity–electrostatic bimorph PVDF material: theoretical, FEM and experimental verification, Microsystem Technologies 25 (8) (2019) 3235–3242.

[5] Vittorio Memmolo, Hassan Elahi, Marco Eugeni, Ernesto Monaco, Fabrizio Ricci, Michele Pasquali, Paolo Gaudenzi, Experimental and numerical investigation of PZT response in composite structures with variable degradation levels, Journal of Materials Engineering and Performance 28 (6) (2019) 3239–3246.

[6] H. Elahi, A. Israr, R.F. Swati, H.M. Khan, A. Tamoor, Stability of piezoelectric material for suspension applications, in: 2017 Fifth International Conference on Aerospace Science & Engineering (ICASE), IEEE, 2017, pp. 1–5.

[7] Hassan Elahi, Marco Eugeni, Paolo Gaudenzi, Electromechanical degradation of piezoelectric patches, in: Holm Altenbach, Erasmo Carrera, Gennady Kulikov (Eds.), Analysis and Modelling of Advanced Structures and Smart Systems, Springer Singapore, Singapore, 2018, pp. 35–44.

[8] Cem Ayyildiz, H. Emre Erdem, Tamer Dirikgil, Oguz Dugenci, Taskin Kocak, Fatih Altun, V. Cagri Gungor, Structure health monitoring using wireless sensor networks on structural elements, Ad Hoc Networks 82 (2019) 68–76.

[9] Shradha Saxena, Rakesh Kumar Dwivedi, Vijay Khare, Simulation study of uncoupled electrical equivalent model of piezoelectric energy harvesting device interfaced with different electrical circuits, in: Advances in Signal Processing and Communication, Springer, 2019, pp. 591–600.

[10] R.F. Swati, Hassan Elahi, L.H. Wen, A.A. Khan, S. Shad, M. Rizwan Mughal, Investigation of tensile and in-plane shear properties of carbon fiber reinforced composites with and without piezoelectric patches for micro-crack propagation using extended finite element method, Microsystem Technologies (2018) 1–10.

16 Piezoelectric Aeroelastic Energy Harvesting

[11] Angela Triplett, D. Dane Quinn, The effect of non-linear piezoelectric coupling on vibration-based energy harvesting, Journal of Intelligent Material Systems and Structures 20 (16) (2009) 1959–1967.

[12] Zubair Butt, Riffat Asim Pasha, Faisal Qayyum, Zeeshan Anjum, Nasir Ahmad, Hassan Elahi, Generation of electrical energy using lead zirconate titanate (PZT-5a) piezoelectric material: analytical, numerical and experimental verifications, Journal of Mechanical Science and Technology 30 (8) (2016) 3553–3558.

[13] Hassan Elahi, Marco Eugeni, Paolo Gaudenzi, A review on mechanisms for piezoelectric-based energy harvesters, Energies 11 (7) (2018) 1850.

[14] Hassan Elahi, Khushboo Munir, Marco Eugeni, Sofiane Atek, Paolo Gaudenzi, Energy harvesting towards self-powered IoT devices, Energies 13 (21) (2020) 5528.

[15] Hassan Elahi, M. Rizwan Mughal, Marco Eugeni, Faisal Qayyum, Asif Israr, Ahsan Ali, Khushboo Munir, Jaan Praks, Paolo Gaudenzi, Characterization and implementation of a piezoelectric energy harvester configuration: analytical, numerical and experimental approach, Integrated Ferroelectrics 212 (1) (2020) 39–60.

[16] Paolo Gaudenzi, Klaus-Jurgen Bathe, An iterative finite element procedure for the analysis of piezoelectric continua, Journal of Intelligent Material Systems and Structures 6 (2) (1995) 266–273.

[17] Paolo Gaudenzi, Rolando Carbonaro, Edoardo Benzi, Control of beam vibrations by means of piezoelectric devices: theory and experiments, Composite Structures 50 (4) (2000) 373–379.

[18] P. Gaudenzi, On the electromechanical response of active composite materials with piezoelectric inclusions, Computers & Structures 65 (2) (1997) 157–168.

[19] Yahia Bouzelata, Erol Kurt, Yunus Uzun, Rachid Chenni, Mitigation of high harmonicity and design of a battery charger for a new piezoelectric wind energy harvester, Sensors and Actuators A: Physical 273 (2018) 72–83.

[20] Mohammed Salim, Dhia Salim, Davannendran Chandran, Hakim S. Aljibori, A.Sh. Kherbeet, Review of nano piezoelectric devices in biomedicine applications, Journal of Intelligent Material Systems and Structures (2018) 1045389X17754272.

[21] Zhi Li, Jinjun Shan, LQG-based synchronization control of Fabry–Perot spectrometer using multiple piezoelectric actuators (PEAs), in: 2016 IEEE International Conference on Information and Automation (ICIA), IEEE, 2016, pp. 448–453.

[22] Demi Ai, Hongping Zhu, Hui Luo, Chao Wang, Mechanical impedance based embedded piezoelectric transducer for reinforced concrete structural impact damage detection: a comparative study, Construction and Building Materials 165 (2018) 472–483.

[23] Hassan Elahi, Zubair Butt, Marco Eugnei, Paolo Gaudenzi, Asif Israr, Effects of variable resistance on smart structures of cubic reconnaissance satellites in various thermal and frequency shocking conditions, Journal of Mechanical Science and Technology 31 (9) (2017) 4151–4157.

[24] Rishikesh Pandey, Dipak Kumar Khatua, Shekhar Tyagi, Mulualem Abebe, Bastola Narayan, Vasant Sathe, Rajeev Ranjan, Length-scale dependent average structures, piezoelectricity enhancement and depolarization mechanisms in a non-MPB high-performance piezoelectric alloy system $PbTiO_3$-$Bi(Zr_{1/2}Ni_{1/2})O_3$, arXiv preprint, arXiv:1801.03310, 2018.

[25] Alper Erturk, Ghislain Delporte, Underwater thrust and power generation using flexible piezoelectric composites: an experimental investigation toward self-powered swimmer-sensor platforms, Smart Materials and Structures 20 (12) (2011) 125013.

[26] P. Muralt, M. Marzencki, B. Belgacem, F. Calame, S. Basrour, Vibration energy harvesting with PZT micro device, Procedia Chemistry 1 (1) (2009) 1191–1194.

[27] Mahidur R. Sarker, Sawal H.M. Ali, Masuri Othman, Md. Shabiul Islam, Designing a low voltage energy harvesting circuits for rectified storage voltage using vibrat-

ing piezoelectric, in: 2011 IEEE Student Conference on Research and Development (SCOReD), IEEE, 2011, pp. 343–346.

[28] Hassan Elahi, Marco Eugeni, Paolo Gaudenzi, Madiha Gul, Raees Fida Swati, Piezoelectric thermo electromechanical energy harvester for reconnaissance satellite structure, Microsystem Technologies (2018) 1–8.

[29] Alireza Keshmiri, Xiaowei Deng, Nan Wu, New energy harvester with embedded piezoelectric stacks, Composites Part B: Engineering 163 (2019) 303–313.

[30] Hassan Elahi, Khushboo Munir, Marco Eugeni, Paolo Gaudenzi, Reliability risk analysis for the aeroelastic piezoelectric energy harvesters, Integrated Ferroelectrics 212 (1) (2020) 156–169.

[31] Hassan Elahi, Marco Eugeni, Luca Lampani, Paolo Gaudenzi, Modeling and design of a piezoelectric nonlinear aeroelastic energy harvester, Integrated Ferroelectrics 211 (1) (2020) 132–151.

[32] Hassan Elahi, Marco Eugeni, Federico Fune, Luca Lampani, Franco Mastroddi, Giovanni Paolo Romano, Paolo Gaudenzi, Performance evaluation of a piezoelectric energy harvester based on flag-flutter, Micromachines 11 (10) (2020) 933.

[33] Hassan Elahi, The investigation on structural health monitoring of aerospace structures via piezoelectric aeroelastic energy harvesting, Microsystem Technologies (2020) 1–9.

[34] Marco Eugeni, Hassan Elahi, Federico Fune, Luca Lampani, Franco Mastroddi, Giovanni Paolo Romano, Paolo Gaudenzi, Numerical and experimental investigation of piezoelectric energy harvester based on flag-flutter, Aerospace Science and Technology 97 (2020) 105634.

[35] Marco Eugeni, Hassan Elahi, Federico Fune, Luca Lampani, Franco Mastroddi, Giovanni Paolo Romano, Paolo Gaudenzi, Experimental evaluation of piezoelectric energy harvester based on flag-flutter, in: Conference of the Italian Association of Theoretical and Applied Mechanics, Springer, 2019, pp. 807–816.

[36] Hassan Elahi, Marco Eugeni, Paolo Gaudenzi, Faisal Qayyum, Raees Fida Swati, Hayat Muhammad Khan, Response of piezoelectric materials on thermomechanical shocking and electrical shocking for aerospace applications, Microsystem Technologies (2018) 1–8.

[37] B. Muruganantham, R. Gnanadass, N.P. Padhy, Challenges with renewable energy sources and storage in practical distribution systems, Renewable and Sustainable Energy Reviews 73 (2017) 125–134.

[38] Minsu Choi, Jaeyoung Park, Jaichan Lee, Design of piezoelectric actuator for braille module by finite element method, Journal of Nanoscience and Nanotechnology 19 (3) (2019) 1308–1314.

[39] Henno Allik, Thomas Hughes Jr., Finite element method for piezoelectric vibration, International Journal for Numerical Methods in Engineering 2 (2) (1970) 151–157.

[40] Wen Tung Chien, Chih Jen Yang, Yu Tang Yen, et al., Coupled-field analysis of piezoelectric beam actuator using FEM, Sensors and Actuators A: Physical 118 (1) (2005) 171–176.

[41] Geffrey K. Ottman, Heath F. Hofmann, Archin C. Bhatt, George A. Lesieutre, Adaptive piezoelectric energy harvesting circuit for wireless remote power supply, IEEE Transactions on Power Electronics 17 (5) (2002) 669–676.

[42] Lianxi Liu, Yanbo Pang, Wenzhi Yuan, Zhangming Zhu, Yintang Yang, A self-powered piezoelectric energy harvesting interface circuit with efficiency-enhanced p-SSHI rectifier, 2018.

[43] G. Melilli, D. Lairez, D. Gorse, E. Garcia-Caurel, A. Peinado, O. Cavani, B. Boizot, M-C. Clochard, Conservation of the piezoelectric response of PVDF films under irradiation, Radiation Physics and Chemistry 142 (2018) 54–59.

[44] Feng Liang, Degang Zhao, Desheng Jiang, Zongshun Liu, Jianjun Zhu, Ping Chen, Jing Yang, Wei Liu, Shuangtao Liu, Yao Xing, et al., Improvement of slope efficiency

of GaN-based blue laser diodes by using asymmetric MQW and $In_xGa_{1-x}N$ lower waveguide, Journal of Alloys and Compounds 731 (2018) 243–247.

[45] Mahmoud Al Ahmad, Areen Allataifeh, Piezoelectric-based energy harvesting for smart city application, in: Information Innovation Technology in Smart Cities, Springer, 2018, pp. 343–356.

[46] Hyeon Lee, Nathan Sharpes, Hichem Abdelmoula, Abdessattar Abdelkefi, Shashank Priya, Higher power generation from torsion-dominant mode in a zigzag shaped two-dimensional energy harvester, Applied Energy 216 (2018) 494–503.

[47] Rashid Naseer, Abdessattar Abdelkefi, Huliang Dai, Lin Wang, Characteristics and comparative analysis of monostable and bistable piezomagnetoelastic energy harvesters under vortex-induced vibrations, in: 2018 AIAA/ASCE/AHS/ASC Structures, Structural Dynamics, and Materials Conference, 2018, p. 1959.

[48] H.L. Dai, Y.W. Yang, A. Abdelkefi, L. Wang, Nonlinear analysis and characteristics of inductive galloping energy harvesters, Communications in Nonlinear Science and Numerical Simulation 59 (2018) 580–591.

[49] Fei Fang, Guanghui Xia, Jianguo Wang, Nonlinear dynamic analysis of cantilevered piezoelectric energy harvesters under simultaneous parametric and external excitations, Acta Mechanica Sinica (2018) 1–17.

[50] Hong-Jun Xiang, Zhiwei Zhang, Zhifei Shi, Hong Li, Reduced-order modeling of piezoelectric energy harvesters with nonlinear circuits under complex conditions, in: Smart Materials and Structures, 2018.

[51] Zhengqiu Xie, C.A. Kitio Kwuimy, Zhiguo Wang, Wenbin Huang, A piezoelectric energy harvester for broadband rotational excitation using buckled beam, AIP Advances 8 (1) (2018) 015125.

[52] Hai Tao Li, Wei Yang Qin, Jean Zu, Zhengbao Yang, Modeling and experimental validation of a buckled compressive-mode piezoelectric energy harvester, Nonlinear Dynamics (2018) 1–20.

[53] Sijun Du, Yu Jia, Ashwin A. Seshia, Piezoelectric vibration energy harvesting: a connection configuration scheme to increase operational range and output power, Journal of Intelligent Material Systems and Structures 28 (14) (2017) 1905–1915.

[54] Xiang Wang, Changsong Chen, Na Wang, Haisheng San, Yuxi Yu, Einar Halvorsen, Xuyuan Chen, A frequency and bandwidth tunable piezoelectric vibration energy harvester using multiple nonlinear techniques, Applied Energy 190 (2017) 368–375.

[55] Guangqing Wang, Wei-Hsin Liao, Binqiang Yang, Xuebao Wang, Wentan Xu, Xiuling Li, Dynamic and energetic characteristics of a bistable piezoelectric vibration energy harvester with an elastic magnifier, Mechanical Systems and Signal Processing 105 (2018) 427–446.

[56] Tien Dat Phan, Patrick Springer, Robert Liebich, Numerical investigation of an elastomer-piezo-adaptive blade for active flow control of a nonsteady flow field using fluid–structure interaction simulations, Journal of Turbomachinery 139 (9) (2017) 091004.

[57] Mohammad Ghalambaz, Esmail Jamesahar, Muneer A. Ismael, Ali J. Chamkha, Fluid–structure interaction study of natural convection heat transfer over a flexible oscillating fin in a square cavity, International Journal of Thermal Sciences 111 (2017) 256–273.

[58] A. Farajpour, A. Rastgoo, M. Mohammadi, Vibration, buckling and smart control of microtubules using piezoelectric nanoshells under electric voltage in thermal environment, Physica B: Condensed Matter 509 (2017) 100–114.

[59] Reza Abdolvand, Hedy Fatemi, Sina Moradian, Quality factor and coupling in piezoelectric MEMS resonators, in: Piezoelectric MEMS Resonators, Springer, 2017, pp. 133–152.

[60] Muhammad Usman Khan, Zubair Butt, Hassan Elahi, Waqas Asghar, Zulkarnain Abbas, Muhammad Shoaib, M. Anser Bashir, Deflection of coupled elasticity–electrostatic

bimorph PVDF material: theoretical, FEM and experimental verification, Microsystem Technologies (2018) 1–8.

[61] Hassan Elahi, Khushboo Munir, Marco Eugeni, Muneeb Abrar, Asif Khan, Adeel Arshad, Paolo Gaudenzi, A review on applications of piezoelectric materials in aerospace industry, Integrated Ferroelectrics 211 (1) (2020) 25–44.

[62] Hassan Elahi, Marco Eugeni, Paolo Gaudenzi, Madiha Gul, Raees Fida Swati, Piezoelectric thermo electromechanical energy harvester for reconnaissance satellite structure, Microsystem Technologies 25 (2) (2019) 665–672.

CHAPTER 2

Smart structures

Contents

2.1.	Introduction	21
2.2.	Piezoelectric smart structures	22
2.3.	Shape memory alloys	22
	2.3.1 Electrorheological (ER) fluids	23
	2.3.2 Magnetorheological (MR) fluids	23
2.4.	Piezoelectric sensors	24
2.5.	Sandwich structures	26
2.6.	Piezothermoelastic materials	27
2.7.	Piezothermoelastic response governing equations	28
2.8.	An example of smart structures application: machine learning based damage assessment	30
	2.8.1 Laminated composites damage assessment with machine learning	31
References		34

2.1 Introduction

If there are some changes in the internal or external environment such as failure/damage or the changes in loads or the shape then a smart structure can effectively respond to these changes [1]. In addition, actuators made of smart materials are incorporated, which enables alternation in the system's response, as well as in the characteristics of the system such as damping/stiffness or shape/strain [2,3]. Therefore there are four main parts of smart structures: power conditioning electronics, sensors, actuators, and control strategies. There are different types of sensor and actuator which are being used for different applications, for example, electrostrictive materials, piezoelectric materials, electro and magnetorheological fluids, shape memory alloys, magnetostrictive materials, and fiber optics. Using the embedding or surface bonding, it is possible to integrate them with the main load-carrying structures without affecting the system's mass or structural stiffness [4,5].

Physical system applications of smart structures include actively controlling the noise, damping, vibration, stress distribution, aeroelastic stability, and shape change. They have a range of applications from systems in space to the automotive, rotary-wing aircraft, civil structures, fixed-wing, medical systems, and machine tools. Many of the early implementations of the

Piezoelectric Aeroelastic Energy Harvesting
https://doi.org/10.1016/B978-0-12-823968-1.00011-8

smart-structure methodology were motivated by space applications such as large flexible space structure's shape control and vibration, but now broader applications for aeronautical and other systems are envisaged. Surface-bonded or embedded smart actuators on the blade of a helicopter or on the wing of an airplane, as an example, can induce camber change/airfoil twist which in turn causes lift distribution variation and helps in controlling dynamic and static aeroelastic problems [6–8].

2.2 Piezoelectric smart structures

The most famous smart materials are made of piezoelectric (PZT) materials. An applied electric field can cause strain (deformation) across the PZT materials, or the other way around, it can be said that the voltage is produced because of strain and therefore they can be used both as sensors and actuators. These materials under the applied field generate very low strain but a wide actuation frequency range is covered. Piezoelectric materials are bipolar and relatively linear but they exhibit hysteresis. Most popular piezoceramics are like thin sheets which can be embedded or attached in composite structures. On the surface, these sheets cause the isotropic strains, and across the thickness they can cause non-Poisson strain. In the form of stacks, the electrostrictive and piezoelectric materials are available, and they are made up of several layers of assembled electrodes and materials. Large forces are generated by these stacks with small displacements in the normal direction to the bottom and top surfaces. There are bending actuators or bimorphs commercially available, where two piezoceramic layers are stacked with a thin shim in-between them. A bending action is created if on these two sheets opposite polarity is applied. In comparison to a single piezoelement, the bimorphs cause small force and large displacement.

2.3 Shape memory alloys

Shape memory alloys (SMAs) seem desirable as actuators due to the likelihood of achieving large displacements and excitation forces. At the specific temperature, these materials show a phase transformation. When deformed plastically at low temperatures, these alloys restore their original undeformed state if their temperature increases above the transformation temperature. It is a reversible process. A large change is seen in the elasticity modulus when heated above the phase transition temperature; this characteristic of SMA is well known. Nitinol is the most commonly used SMA material which is a nickel–titanium alloy that is available with different diameters in the form of wires. It is possible to carry out the SMA heating externally with the help of coils, as well as internally (electric resistance), but there is a very slow response, probably less than 1 Hz. Forced convective or conductive material cooling can speed up the response. There is a similarity between piezoelectric and electrostrictive materials as they have the same strain capability. These are monopolar, temperature-sensitive, with nonlinear relation between induced strain and applied field. They exhibit negligible hysteresis.

When a magnetic field is applied to magnetostrictive materials such as Terfenol-D, they elongate. These materials exhibit hysteresis, are nonlinear, and monopolar. Over a wide range of frequencies, these materials can moderate forces and can generate low strains. These actuators are also bulky due to the coil and the magnetic return path.

2.3.1 Electrorheological (ER) fluids

Electrorheological (ER) fluids contain a fine dielectric particles' suspension in an insulating fluid, exhibiting, in the presence of a large applied electric field up to 4 kV/mm, a controlled rheological behavior. A shear loss factor change can be seen when an electric field is applied; this helps in altering the system's damping.

2.3.2 Magnetorheological (MR) fluids

Magnetorheological (MR) fluids are composed of a ferrous particles' suspension in the fluid, exhibiting the sheer loss factor change influenced by magnetic fields (large to moderate currents with low fields). Just like ER fluids, the MR fluids augment the system's damping. There is an increase in the popularity of fiber optics due to its ability to get embedded easily in the composite structures with a small effect on the integrity of the structure and with the potential ability of multiplexing.

Smart structures are feasible because of the following:
1. Commercial availability of smart materials;
2. In laminated structures, it is easy to embed the devices;
3. Material coupling can be exploited to produce desirable electrical and mechanical properties;
4. At a small price, they potentially increase performance (weight penalty);
5. Advances in sensor technology, information processing, and microelectronics.

24 Piezoelectric Aeroelastic Energy Harvesting

Hardware, control methodology, sensors, and actuators are the key elements for smart structure technologies.

2.4 Piezoelectric sensors

When stress is applied mechanically on the piezoelectric material, an electric displacement is generated, which is the direct piezoelectric effect, and can be used for structural deformation. The majority of applications depend on either the voltage or rate of voltage change produced either by the sensor or by the spectrum of frequencies of the generated signal by sensors. The higher-frequency noise rejection and superior signal-to-noise ratio of piezoelectric sensors give them the biggest advantage over the conventional foil strain gauges. In piezoelectric sensors, less signal conditioning is required for applications where a low level of strains is involved and they are less sensitive to noise. Over a large bandwidth, strain sensitivity, compactness, and embeddability ease.

Piezofilms, i.e., PVDFs, are the most used sensors due to the low stiffness. For some applications, piezoceramic (PZT) sensors are used. For both actuation and sensing, the piezoceramics can be used under the collocated control strategies [9–12]. The PZT sensors exhibit brittleness, high Young's modulus, and low tensile strength [13–18]. Also under the conditions of high stress, they may suffer depolarization, and from creep with DC field and at high strains, they suffer from linearity. Nevertheless, a few researchers made use of the piezoceramic sheet sensors under the structural systems which are controllable [19,20] and in the applications for monitoring health [21]. Different PZT sheet sensors were evaluated comparatively by Giurgiutiu and Zagrai [22]. The behavior of PVDF is orthotropic piezoelectrically due to stretching in materials during manufacturing, but for small strains it is mechanically isotropic.

The equations of the direct effect are

$$
\begin{Bmatrix} D_1 \\ D_2 \\ D_3 \end{Bmatrix} = \begin{bmatrix} 0 & 0 & 0 & 0 & d_{15} & 0 \\ 0 & 0 & 0 & d_{25} & 0 & 0 \\ d_{31} & d_{32} & d_{33} & 0 & 0 & 0 \end{bmatrix} \begin{Bmatrix} \sigma_1 \\ \sigma_2 \\ \sigma_3 \\ \tau_{23} \\ \tau_{31} \\ \tau_{12} \end{Bmatrix}
$$

$$+ \begin{bmatrix} e_{11}^{\sigma} & 0 & 0 \\ 0 & e_{22}^{\sigma} & 0 \\ 0 & 0 & e_{33}^{\sigma} \end{bmatrix} \begin{Bmatrix} E_1 \\ E_2 \\ E_3 \end{Bmatrix} + \begin{bmatrix} a_1 \\ a_2 \\ a_3 \end{bmatrix} \Delta T. \qquad (2.1)$$

The equations of a sensor are direct effect-based. An electric field is generated when a sensor is exposed to a stress field. Monolithic PZT sensors are transversely isotropic, and so $d_{32} = d_{31}$ and $d_{25} = d_{15}$. When a zero electric field and no thermal strain are considered, we obtain

$$\begin{Bmatrix} D_1 \\ D_2 \\ D_3 \end{Bmatrix} = \begin{bmatrix} 0 & 0 & 0 & 0 & d_{15} & 0 \\ 0 & 0 & 0 & d_{25} & 0 & 0 \\ d_{31} & d_{32} & d_{33} & 0 & 0 & 0 \end{bmatrix} \begin{Bmatrix} \sigma_1 \\ \sigma_2 \\ \sigma_3 \\ \tau_{23} \\ \tau_{31} \\ \tau_{12} \end{Bmatrix}. \qquad (2.2)$$

a charge q can be captured on the boundaries of the piezoelectric body by means of conductive plates (electrodes) so that:

$$q = \int_s D_i n_i ds \qquad (2.3)$$

where n_i represents a normal vector to the surface s.

Moreover, the V_c voltage and q charge are related to the sensor's capacitance C_p as

$$V_c = \frac{q}{C_p}. \qquad (2.4)$$

Strain and stress can be measured once the voltage is known. Now consider a two-faced sheet sensor coated with thin layers of the electrode. The capacitance in the uniaxial stress field case is then given as

$$C_p = \frac{e_{33}^{\sigma} l_c b_c}{t_c}, \qquad (2.5)$$

where l_c is the length, b_c is the width, and t_c is the thickness of the sensor.

Now if we assume only unidirectional strain then

$$V_c = \frac{d_{31} Y_c b_c}{C_p} \int_{l_c} \epsilon_1 dx, \qquad (2.6)$$

where sensor's Young's modulus is denoted by Y_c and ϵ_1 is averaged over gauge length.

The strain is then given as

$$\epsilon_1 = \frac{V_c C_p}{d_{31} Y_c l_c b_c}.$$

(2.7)

Unidirectional strain is assumed for this relationship and so there is no strain loss in the bond layer. If Poisson's effect is considered then this relation is reduced to

$$\epsilon_1 = \frac{V_c C_p}{d_{31}[1 - \nu(\frac{d_{32}}{d_{31}})]Y_c l_c b_c}.$$

(2.8)

Here ν is Poisson's ratio. The transverse sensitivity for the conventional foil gauges is closer to zero, which is neglected normally. In general, by using just one piezoelectric sensor, separating the principle strain is not possible. Obtaining the longitudinal strain with the help of one sensor is not possible unless the prior transverse strain is known.

When the uniform stain beam theory is assumed, the bond layer effect can be determined. The effective width and length of the sensor reduce because of the effect of shear lag in the bond layer:

$$\epsilon_1 = \frac{V_c C_p}{d_{31}}[1 - \nu(\frac{d_{32}}{d_{31}})]Y_c l_{c\text{eff}} b_{c\text{eff}}.$$

(2.9)

These effective widths and lengths are functions of sensor and bond–layer characteristics. A procedure for the determination of these values is given in [23]. Shear lag loss is smaller in the PZT sensor than in the PVDF sensor because of the small stiffness and thickness.

2.5 Sandwich structures

Considering the quasi–metal–derived design, the sandwich concept has more advantages. There is a reduction in the complexity of maintenance, manufacturing, and analysis due to integral simple stiffening which in turn cuts down the cost of the total life cycle of a structure. When sandwich structures are introduced to the fuselage shells, this results in the reduction of the noise level in the cabin, due to the core material's high damping properties, and this increases the comfort of the passenger.

With the support of the German government in 1999, a CFRP fuselage project was initiated by Airbus [24]. New sandwich concepts were formed after a wide screening of technologies named Venable Shear Core (VeSCo) and Stringer outside, Frames inside (SoFi). Both of them use innovative core materials and are integrated double-shell structures, i.e., a folded core which is like the origami construction [25] and carbon pin reinforced foam [25], which enables implementing more functions into fewer parts. This fuselage concept was also evaluated by Boeing and NASA, and they concluded that up to 41% of offered cost and 29% of offered weight can be saved in comparison with baseline aluminum design, whereas only 23% fuselage is achieved with composite skin-stringer [26]. According to NASA, the composite sandwich structures are considered as a promising structural concept for the primary structure of the future vehicles of space transportation [27].

2.6 Piezothermoelastic materials

Piezoelectric materials because of their special characteristics can effectively function as distributed sensors and structural response controlling actuators. The measurement of the direct piezoelectric effect (induced electric potential difference) can help in determining the thermally or mechanically induced disturbances of the sensor. On the other hand, in applications of actuators, the introduction of the converse piezoelectric effect (appropriate electric potential) can help to control the stress or deformation. When advanced composite materials and the piezoelectric materials are integrated together, there exists a potential that forms high-stiffness, high-strength, and light-weighted structures with the capability of self-controlling and self-monitoring [28–31].

Boundary conditions and the equations which govern the behavior of an electrically polarizable, heat-conducting, and finitely deformable medium with an electric field interaction were derived by Tiersten [32]. Later on, Mindlin [33] deduced equations for the vibrations of the piezoelectric plate, including a coupling between the temperature, deformation, and electric field. The authors of [33] presented a uniqueness theorem for the edge and face conditions in order to have a unique solution corresponding to 2D equations.

28 Piezoelectric Aeroelastic Energy Harvesting

2.7 Piezothermoelastic response governing equations

The linear response of a piezothermoelastic medium described by the governing equations' differential form, in curvilinear coordinates, is written as follows:

(Electrostatics)

$$D_{i,i} = 0, \tag{2.10}$$

(Mechanical equilibrium)

$$\sigma_{ij,j} + f_i = \rho \ddot{u}_i, \tag{2.11}$$

(Heat conduction)

$$E_\zeta = -(\phi_1)_{,\zeta}, \quad E_z = -(\phi_0 + \phi_1)_{,z}, \tag{2.12}$$

$$u_\zeta = (\phi_1 + \phi_2 + \sum_{i=1}^{3} m_i \psi_i)_{,\zeta}, \tag{2.13}$$

$$u_z = (\phi_0 + k_1 \phi_1 + j\phi_2 + \sum_{i=1}^{3} m_i \psi_i)_{,z}. \tag{2.14}$$

The potential functions for $T_0(z, t)$ temperature field, i.e., Φ_0 and ϕ_0, are

$$\Phi_0 = \int \gamma_2 T_0(z, t) dz + C_1, \tag{2.15}$$

$$\phi_{0,z} = \int \gamma_1 T_0(z, t) dz + C_2. \tag{2.16}$$

The electric potential ϕ_1 for $T_1(\zeta, z, t)$ is given by

$$\Phi_1 = \chi + (\sum_{i=1}^{3} \eta_i \psi_i)_{,z}. \tag{2.17}$$

The piezoelectric function χ and piezothermoelastic potentials ψ_i ($i = 1, 2, 3$) in case of piezothermoelastic potentials ϕ_i ($i = 1, 2$) are governed by the equations:

$$(\Delta_\zeta + \mu_1 \frac{\partial^2}{\partial z^2})(\Delta_\zeta + \mu_2 \frac{\partial^2}{\partial z^2})(\Delta_\zeta + \mu_3 \frac{\partial^2}{\partial z^2})\phi_1$$
$$= d_2 \Delta_\zeta T_{1,zz} + d_1 \Delta_\zeta \Delta_\zeta T_1 + d_1 \Delta_\zeta T_{1,zz} + d_0 T_{1,zzz}, \tag{2.18}$$

$$\Delta_\zeta \psi_i + \mu_i \psi_{i,zz} = 0, \tag{2.19}$$

$$\phi_{2,zz} = \frac{1}{\xi_1}(\Delta_\zeta \phi_1 + \nu_1 \phi_{1,zz} - \delta_1 T_1), \tag{2.20}$$

$$\chi_{,z} = \frac{1}{\xi_2}(\Delta_\zeta \phi_2 + \nu_2 \phi_{2,zz} - \delta_2 T_1), \tag{2.21}$$

$$\Delta_\zeta = \begin{cases} \frac{\partial^2}{\partial x^2}, & \zeta = x, \\ \frac{\partial^2}{\partial r^2} + r^{-1}\frac{\partial}{\partial r}, & \zeta = r. \end{cases} \tag{2.22}$$

The quantities in the above equations, l_i, k_1, j, m_i, ξ_i, d_i, n_i, ν_i, γ_i, and μ_i, depend on the material properties, as given by the authors in [34] for $\zeta = x$, and in [35] for $\zeta = r$.

With the deformation of piezoelectric potential, an electric potential is generated, providing a sensing mechanism for thermomechanical disturbances. The main purpose of piezoelectric element utilization is the unknown loading's prediction or energy harvesting [36–39]. Many authors investigated the corresponding responses of the piezothermoelasticity from induced electric potential measurements [40–45]. The authors of [43] solved an inversion problem, while the researchers in [46] considered the problem of a disk subjected to the time-varying, spatially uniform ambient surface temperature. The axisymmetric transient response case was considered in [44]. The authors presented a procedure of finite difference for the determination of the transient, radially-varying ambient temperature on a circular disk's surface which was based on the electric potential difference induced across the thickness of the disk. A thermal loading causes the electric potential difference given by

$$V(r, t) = [\Phi]_{z=b} - [\Phi]_{z=-b} = -V_0(1 - 2f\frac{r^2}{a^2} + f\frac{r^4}{a^4})(1 - e^{\frac{-kt}{a^2}}), \tag{2.23}$$

where f is a specified parameter and V_0 is a constant. The radial distribution of the dimensionless electric potential difference is given by

$$[\bar{\Phi}]_{\bar{z}=\bar{b}} - [\bar{\Phi}]_{\bar{z}=-\bar{b}}. \tag{2.24}$$

For the disk ratio of thickness-to-diameter $\bar{b} = \frac{b}{a} = 0.1$, at various $\bar{t}$ times, the corresponding parameter value is $f = 1$.

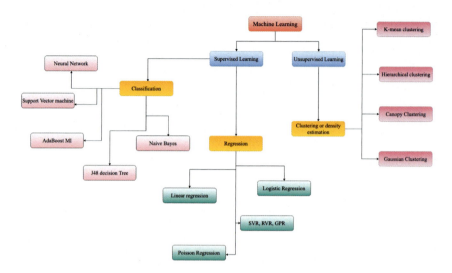

Figure 2.1 Machine learning, its objectives, and commonly used algorithms.

2.8 An example of smart structures application: machine learning based damage assessment

Literature shows that discriminative feature extraction for pristine and damaged composite structures is performed in the time, impedance, frequency, modal analysis, and time–frequency domains [47–52]. In the domains the discriminative features can be extracted either from high-frequency with guided waves of low-wavelength or from low-frequency with guided waves of high-wavelength, or ultrasonic and acoustic waves [53–56]. A few examples of the extraction of these discriminative features from high-wavelength low-frequency structural vibration response include the damping capacity, strain mode shapes, natural frequency, mode shape curvature, frequency response shift from baseline, and power spectral density form composite structures [52,57–62]. The features or the descriptors for sensing the damage, which are commonly employed for the high-frequency and low-wavelength guided or acoustic waves, are the duration, peak amplitude, counts, rise time, energy, counts to peak, average frequency, and peak frequency [63–66].

Machine learning technology has become a powerful tool in the regression clustering and classification of discriminatory characteristics by drawing hyperplane's boundaries of complex decisions [67–70]. However, given the previous discussion, a superset of discriminative features cannot be established which would distinguish all types of damage modes, such as fiber fracture, delamination, voids, matrix cracks, etc., in the composite structures, and at the same time quantify, identify, and localize the specific damage type withe same discriminative feature set. Additionally, one machine learning technique can define a specific type of damage effectively but can fail for others. This section is thus intended to examine the damage-sensitive characteristics of different forms of damage caused by composite structures in detail. Confusion matrix for the third level Daubechies-4 MLP classification network is expressed in Table 2.1.

Table 2.1 Confusion matrix for the third level Daubechies-4 MLP classification network [71].

True Class	Predicted Class			
	UD	DR 1	DR 2	DR 3
UD	40	0	0	0
DR 1	1	189	7	3
DR 2	0	10	88	2
DR 3	4	15	7	174

2.8.1 Laminated composites damage assessment with machine learning

Machine learning techniques are used in three situations, according to Cherkassky and Mulier [72]:

1. Classification – developing a computational model by combining measured quantity vectors and discrete labels. After training the model can be used for the prediction of new measurements.
2. Regression – model development based on the training samples, with the association of continuous value targets.
3. Density estimation – without any target values or the output, estimating the probability density function of the measured data samples.

Supervised methods of learning are generally used to address classification and regression problems, while the calculation of density is carried out by unsupervised learning. Supervised learning is builds on observational evidence of the computational models/rules. In supervised learning, mapping functions are learned on the basis of knowledge of the inputs to known outputs through a learned and error-corrected approach. Then these learned functions are used for the prediction of the new inputs/observations, whereas in unsupervised learning the underlying pattern or structure is learned within the measurement dataset. Unlike supervised learning, algorithms with unsupervised learning do not require any output or observation label. In Fig. 2.1, machine learning types and their objectives are shown.

The origins of ANN can be traced back to the late 1800s when scientific attempts were made to recreate the function of the human brain. However, Ghaboussi et al. [73] were the first to employ ANN for the detection of damages. Literature shows the successful implementation of ANN for the fault detection in the bearing [74], induction motors [75], stochastic nonlinear system identification [76], gear fault detection and diagnostics [77], and ship design [78].

Localization and quantification of delamination in smart composite laminates were achieved by using the back-propagation neural network by Islam et al. [71]. First, five frequencies of the modal were used for the training of the neural network as damage-sensitive features, and unseen delamination cases at a different location were used for the testing of the trained network. In a smart composite beam, delamination size was assessed with a feedforward back-propagation neural network by Okafor et al. [79]. Here they trained the network on the first four frequencies of the modal and dimensionless delamination sizes. Testing of the network was done on the new delamination cases where the first four normalized frequencies of the test cases were used. A successful prediction was done by the network with the dimensionless delamination size between 0.22 and 0.82, but it did not perform well with the sizes below 0.08.

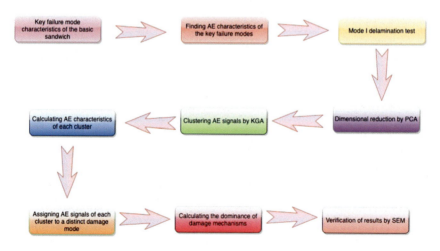

Figure 2.2 A concise explanation of the damage classification procedure [80].

A combination of Levenberg–Marquardt algorithm with neural network and generalization method was used for a reliable and accurate localization of the damage of low-velocity impact in smart composite laminates. Convolutional neural networks were combined with the X-ray computed tomography by Sammons et al. [81] for the quantification and identification of the delamination in a carbon fiber reinforced polymer composite. Based on the central pixel of the patches with size 201μ, class labels were assigned. The network performed well on the small-sized delaminations, but it underperformed on the large sizes.

Table 2.2 Central characteristic of three classes obtained from FCM classification for specimen A1 [82].

Parameter of Signal	Average Freq. (kHz)	Peak Amplitude (dB)	Count
1st Class	139.95	49.23	10.18
2nd Class	256.52	53.47	12.61
3rd Class	412.2	55.12	11.74

Classification and identification of the failure modes in glass fiber reinforced plastics were carried out by Bar et al. [83] with artificial neural networks from AE signals of static tensile tests measured with a surface mounted PVDF film. The class information was generated by using Kohonen self-organizing feature map (KSOM) for different failure modes (i.e., fiber-matrix, matrix crack, fiber fracture, debonding). The class information obtained from KSOM was then used for the multilayer perception (MLP) training. Different failure modes were successfully discriminated by this technique, irrespective of parameter overlapping of AE signals associated with those modes.

A genetic algorithm was combined with the k-means algorithm for the clustering of the mechanisms of various damage in mode I delamination test of a sandwich composite by Pashmforoush et al. [80]. In Fig. 2.2, the procedure employed for unsupervised damage classification can be seen.

Table 2.3 Average dependency percentage of signals for three classes [82].

Experimental Conditions	Dependency on 1st Class	Dependency on 2nd Class	Dependency on 3rd Class
Specimen A1	54%	37%	9%
Specimen A2	11%	24%	65%

The relationship between unidirectional $[0]_s$ glass /epoxy composites and damage mechanisms of woven $[0, 90]_s$ and the AE signals of quasi-static three-point bending test by the wavelet packet transform (WPT) and Fuzzy C-mean clustering was studied by Fotouhi et al. [82]. The descriptors used for the various fracture mechanisms, such as fiber breakage, debonding, and matrix cracking, were the counts, peak amplitude, and the average frequency of the acoustic signal. The results showed that FCM successfully classifies the three modes of damage of the AE, whereas fracture mechanisms were different for the unidirectional (A2) samples and woven (A1), as can be seen from Tables 2.2 and 2.3.

References

[1] Paolo Gaudenzi, Smart Structures: Physical Behaviour, Mathematical Modelling and Applications, John Wiley & Sons, 2009.

[2] Hassan Elahi, Khushboo Munir, Marco Eugeni, Paolo Gaudenzi, Reliability risk analysis for the aeroelastic piezoelectric energy harvesters, Integrated Ferroelectrics 212 (1) (2020) 156–169.

[3] Hassan Elahi, Marco Eugeni, Luca Lampani, Paolo Gaudenzi, Modeling and design of a piezoelectric nonlinear aeroelastic energy harvester, Integrated Ferroelectrics 211 (1) (2020) 132–151.

[4] Hassan Elahi, The investigation on structural health monitoring of aerospace structures via piezoelectric aeroelastic energy harvesting, Microsystem Technologies (2020) 1–9.

[5] Hassan Elahi, Marco Eugeni, Federico Fune, Luca Lampani, Franco Mastroddi, Giovanni Paolo Romano, Paolo Gaudenzi, Performance evaluation of a piezoelectric energy harvester based on flag-flutter, Micromachines 11 (10) (2020) 933.

[6] Marco Eugeni, Hassan Elahi, Federico Fune, Luca Lampani, Franco Mastroddi, Giovanni Paolo Romano, Paolo Gaudenzi, Experimental evaluation of piezoelectric energy harvester based on flag-flutter, in: Conference of the Italian Association of Theoretical and Applied Mechanics, Springer, 2019, pp. 807–816.

[7] Hassan Elahi, Marco Eugeni, Paolo Gaudenzi, Design and performance evaluation of a piezoelectric aeroelastic energy harvester based on the limit cycle oscillation phenomenon, Acta Astronautica 157 (2019) 233–240.

[8] Marco Eugeni, Hassan Elahi, Federico Fune, Luca Lampani, Franco Mastroddi, Giovanni Paolo Romano, Paolo Gaudenzi, Numerical and experimental investigation of piezoelectric energy harvester based on flag-flutter, Aerospace Science and Technology 97 (2020) 105634.

[9] Thomas Bailey, James E. Hubbard Jr., Distributed piezoelectric-polymer active vibration control of a cantilever beam, Journal of Guidance, Control, and Dynamics 8 (5) (1985) 605–611.

[10] S. Hanagud, M.W. Obal, A.J. Calise, Optimal vibration control by the use of piezoceramic sensors and actuators, Journal of Guidance, Control, and Dynamics 15 (5) (1992) 1199–1206.

[11] Jeffrey J. Dosch, Daniel J. Inman, Ephrahim Garcia, A self-sensing piezoelectric actuator for collocated control, Journal of Intelligent Material Systems and Structures 3 (1) (1992) 166–185.

[12] E.H. Anderson, N.W. Hagood, Simultaneous piezoelectric sensing/actuation: analysis and application to controlled structures, Journal of Sound and Vibration 174 (5) (1994) 617–639.

[13] Mohsin Ali Marwat, Weigang Ma, Pengyuan Fan, Hassan Elahi, Chanatip Samart, Bo Nan, Hua Tan, David Salamon, Baohua Ye, Haibo Zhang, Ultrahigh en-

ergy density and thermal stability in sandwich-structured nanocomposites with dopamine@Ag@BaTiO$_3$, Energy Storage Materials 31 (2020) 492–504.

[14] Ahsan Ali, Riffat Asim Pasha, Hassan Elahi, Muhammad Abdullah Sheeraz, Saima Bibi, Zain Ul Hassan, Marco Eugeni, Paolo Gaudenzi, Investigation of deformation in bimorph piezoelectric actuator: analytical, numerical and experimental approach, Integrated Ferroelectrics 201 (1) (2019) 94–109.

[15] Muhammad Usman Khan, Zubair Butt, Hassan Elahi, Waqas Asghar, Zulkarnain Abbas, Muhammad Shoaib, M. Anser Bashir, Deflection of coupled elasticity–electrostatic bimorph PVDF material: theoretical, FEM and experimental verification, Microsystem Technologies 25 (8) (2019) 3235–3242.

[16] Vittorio Memmolo, Hassan Elahi, Marco Eugeni, Ernesto Monaco, Fabrizio Ricci, Michele Pasquali, Paolo Gaudenzi, Experimental and numerical investigation of PZT response in composite structures with variable degradation levels, Journal of Materials Engineering and Performance 28 (6) (2019) 3239–3246.

[17] H. Elahi, A. Israr, R.F. Swati, H.M. Khan, A. Tamoor, Stability of piezoelectric material for suspension applications, in: 2017 Fifth International Conference on Aerospace Science & Engineering (ICASE), IEEE, 2017, pp. 1–5.

[18] Hassan Elahi, Marco Eugeni, Paolo Gaudenzi, Electromechanical degradation of piezoelectric patches, in: Analysis and Modelling of Advanced Structures and Smart Systems, Springer, 2018, pp. 35–44.

[19] Jinhao Qiu, Junji Tani, Vibration control of a cylindrical shell using distributed piezoelectric sensors and actuators, Journal of Intelligent Material Systems and Structures 6 (4) (1995) 474–481.

[20] C-K. Lee, T.C. O'Sullivan, Piezoelectric strain rate gages, The Journal of the Acoustical Society of America 90 (2) (1991) 945–953.

[21] Paul D. Samuel, Darryll J. Pines, Health monitoring and damage detection of a rotorcraft planetary geartrain system using piezoelectric sensors, in: Smart Structures and Materials 1997: Smart Structures and Integrated Systems, vol. 3041, International Society for Optics and Photonics, 1997, pp. 44–53.

[22] Victor Giurgiutiu, Andrei N. Zagrai, Characterization of piezoelectric wafer active sensors, Journal of Intelligent Material Systems and Structures 11 (12) (2000) 959–976.

[23] Jayant Sirohi, Inderjit Chopra, Fundamental understanding of piezoelectric strain sensors, Journal of Intelligent Material Systems and Structures 11 (4) (2000) 246–257.

[24] Axel S. Herrmann, Pierre C. Zahlen, Ichwan Zuardy, Sandwich structures technology in commercial aviation, in: Sandwich Structures 7: Advancing with Sandwich Structures and Materials, Springer, 2005, pp. 13–26.

[25] Denis D. Cartie, Norman A. Fleck, The effect of pin reinforcement upon the through-thickness compressive strength of foam-cored sandwich panels, Composites Science and Technology 63 (16) (2003) 2401–2409.

[26] L.B. Ilcewicz, P.J. Smith, C.T. Hanson, T.H. Walker, S.L. Metschan, G.E. Mabson, K.S. Wilden, B.W. Flynn, D.B. Scholz, D.R. Polland, et al., Advanced technology composite fuselage: program overview, 1997.

[27] Thomas S. Gates, Xiaofeng Su, Frank Abdi, Gregory M. Odegard, Helen M. Herring, Facesheet delamination of composite sandwich materials at cryogenic temperatures, Composites Science and Technology 66 (14) (2006) 2423–2435.

[28] Hassan Elahi, Zubair Butt, Marco Eugnei, Paolo Gaudenzi, Asif Israr, Effects of variable resistance on smart structures of cubic reconnaissance satellites in various thermal and frequency shocking conditions, Journal of Mechanical Science and Technology 31 (9) (2017) 4151–4157.

[29] Hassan Elahi, Marco Eugeni, Paolo Gaudenzi, Faisal Qayyum, Raees Fida Swati, Hayat Muhammad Khan, Response of piezoelectric materials on thermomechanical shocking and electrical shocking for aerospace applications, Microsystem Technologies 24 (9) (2018) 3791–3798.

[30] Hassan Elahi, Marco Eugeni, Paolo Gaudenzi, Madiha Gul, Raees Fida Swati, Piezoelectric thermo electromechanical energy harvester for reconnaissance satellite structure, Microsystem Technologies 25 (2) (2019) 665–672.

[31] Hassan Elahi, Khushboo Munir, Marco Eugeni, Muneeb Abrar, Asif Khan, Adeel Arshad, Paolo Gaudenzi, A review on applications of piezoelectric materials in aerospace industry, Integrated Ferroelectrics 211 (1) (2020) 25–44.

[32] H.F. Tiersten, On the nonlinear equations of thermo-electroelasticity, International Journal of Engineering Science 9 (7) (1971) 587–604.

[33] R.D. Mindlin, Equations of high frequency vibrations of thermopiezoelectric crystal plates, International Journal of Solids and Structures 10 (6) (1974) 625–637.

[34] Fumihiro Ashida, Naotake Noda, R. Theodore Tauchert, A two-dimensional piezothermoelastic problem in an orthotropic plate exhibiting crystal class mm2, JSME International Journal Series A, Mechanics and Material Engineering 37 (4) (1994) 334–340.

[35] Ashida Fumihiro, Theodore R. Tauchert, Noda Naotake, Response of a piezothermoelastic plate of crystal class 6 mm subject to axisymmetric heating, International Journal of Engineering Science 31 (3) (1993) 373–384.

[36] Zubair Butt, Riffat Asim Pasha, Faisal Qayyum, Zeeshan Anjum, Nasir Ahmad, Hassan Elahi, Generation of electrical energy using lead zirconate titanate (PZT-5a) piezoelectric material: analytical, numerical and experimental verifications, Journal of Mechanical Science and Technology 30 (8) (2016) 3553–3558.

[37] Hassan Elahi, Marco Eugeni, Paolo Gaudenzi, A review on mechanisms for piezoelectric-based energy harvesters, Energies 11 (7) (2018) 1850.

[38] Hassan Elahi, Khushboo Munir, Marco Eugeni, Sofiane Atek, Paolo Gaudenzi, Energy harvesting towards self-powered IoT devices, Energies 13 (21) (2020) 5528.

[39] Hassan Elahi, M. Rizwan Mughal, Marco Eugeni, Faisal Qayyum, Asif Israr, Ahsan Ali, Khushboo Munir, Jaan Praks, Paolo Gaudenzi, Characterization and implementation of a piezoelectric energy harvester configuration: analytical, numerical and experimental approach, Integrated Ferroelectrics 212 (1) (2020) 39–60.

[40] Fumihiro Ashida, Jeong-Seok Choi, Naotake Noda, An inverse thermoelastic problem in an isotropic plate associated with a piezoelectric ceramic plate, Journal of Thermal Stresses 19 (2) (1996) 153–167.

[41] Fumihiro Ashida, Naotake Noda, R. Theodore Tauchert, Inverse problem of two-dimensional piezothermoelasticity in an orthotropic plate exhibiting crystal class mm2, JSME International Journal Series A, Mechanics and Material Engineering 37 (4) (1994) 341–346.

[42] F. Ashida, N. Noda, T.R. Tauchert, Solution method for two-dimensional piezothermoelastic problem of orthotropic solids, Transactions of the JSME, Series A 59 (560) (1993) 946–950.

[43] Fumihiro Ashida, Theodore R. Tauchert, Temperature determination for a contacting body based on an inverse piezothermoelastic problem, International Journal of Solids and Structures 34 (20) (1997) 2549–2561.

[44] Fumihiro Ashida, Theodore R. Tauchert, A finite difference scheme for inverse transient piezothermoelasticity problems, Journal of Thermal Stresses 21 (3–4) (1998) 271–293.

[45] Theodore R. Tauchert, Fumihiro Ashida, Naotake Noda, Recent developments in piezothermoelasticity: inverse problems relevant to smart structures, JSME International Journal Series A, Solid Mechanics and Material Engineering 42 (4) (1999) 452–458.

[46] Fumihiro Ashida, T.R. Tauchert, An inverse problem for determination of transient surface temperature from piezoelectric sensor measurement, 1998.

[47] W.J. Staszewski, Geoffrey Tomlinson, Christian Boller, Geof Tomlinson, Health Monitoring of Aerospace Structures, Wiley Online Library, 2004.

Smart structures **37**

[48] Fan Wei, Pizhong Qiao, Vibration-based damage identification methods: a review and comparative study, Structural Health Monitoring 10 (1) (2011) 83–111.

[49] Scott W. Doebling, Charles R. Farrar, Michael B. Prime, Daniel W. Shevitz, Damage identification and health monitoring of structural and mechanical systems from changes in their vibration characteristics: a literature review, 1996.

[50] E. Peter Carden, Paul Fanning, Vibration based condition monitoring: a review, Structural Health Monitoring 3 (4) (2004) 355–377.

[51] Benjamin L. Grisso, Daniel M. Peairs, Daniel J. Inman, Impedance-based health monitoring of composites, in: IMAC XXII, Dearborn, MI, Jan. 26–29, 2004.

[52] Bin Huang, Heung Soo Kim, Frequency response analysis of a delaminated smart composite plate, Journal of Intelligent Material Systems and Structures 26 (9) (2015) 1091–1102.

[53] Bin Huang, Bong-Hwan Koh, Heung Soo Kim, PCA-based damage classification of delaminated smart composite structures using improved layerwise theory, Computers & Structures 141 (2014) 26–35.

[54] Casey J. Keulen, M. Yildiz, Afzal Suleman, Damage detection of composite plates by lamb wave ultrasonic tomography with a sparse hexagonal network using damage progression trends, Shock and Vibration (2014).

[55] S. Mahadev Prasad, Krishnan Balasubramaniam, C.V. Krishnamurthy, Structural health monitoring of composite structures using lamb wave tomography, Smart Materials and Structures 13 (5) (2004) N73.

[56] D.G. Aggelis, N-M. Barkoula, T.E. Matikas, A.S. Paipetis, Acoustic structural health monitoring of composite materials: damage identification and evaluation in cross ply laminates using acoustic emission and ultrasonics, Composites Science and Technology 72 (10) (2012) 1127–1133.

[57] K. Alnefaie, Finite element modeling of composite plates with internal delamination, Composite Structures 90 (1) (2009) 21–27.

[58] C. Kyriazoglou, B.H. Le Page, F.J. Guild, Vibration damping for crack detection in composite laminates, Composites Part A: Applied Science and Manufacturing 35 (7–8) (2004) 945–953.

[59] Pizhong Qiao, Kan Lu, Wahyu Lestari, Jialai Wang, Curvature mode shape-based damage detection in composite laminated plates, Composite Structures 80 (3) (2007) 409–428.

[60] Cole S. Hamey, Wahyu Lestari, Pizhong Qiao, Gangbing Song, Experimental damage identification of carbon/epoxy composite beams using curvature mode shapes, Structural Health Monitoring 3 (4) (2004) 333–353.

[61] Bin Huang, Heung Soo Kim, Transient analysis of biocomposite laminates with delamination, Journal of Nanoscience and Nanotechnology 14 (10) (2014) 7432–7438.

[62] Heung Soo Kim, Aditi Chattopadhyay, Anindya Ghoshal, Characterization of delamination effect on composite laminates using a new generalized layerwise approach, Computers & Structures 81 (15) (2003) 1555–1566.

[63] Markus G.R. Sause, A. Gribov, Antony R. Unwin, S. Horn, Pattern recognition approach to identify natural clusters of acoustic emission signals, Pattern Recognition Letters 33 (1) (2012) 17–23.

[64] T.M. Ely, EvK Hill, Longitudinal splitting and fiber breakage characterization in graphite epoxy using acoustic emission data, NDT and E International 2 (30) (1997) 109.

[65] M. Suzuki, H. Nakanishi, M. Iwamoto, E. Jinen, Application of static fracture mechanisms to fatigue fracture behavior of class A-SMC composite, in: 4th Japan–U.S. Conference on Composite Materials, Washington, DC, 1989, pp. 297–306.

[66] R. Gutkin, C.J. Green, S. Vangrattanachai, S.T. Pinho, P. Robinson, P.T. Curtis, On acoustic emission for failure investigation in CFRP: pattern recognition and peak frequency analyses, Mechanical Systems and Signal Processing 25 (4) (2011) 1393–1407.

[67] John Joseph Valletta, Colin Torney, Michael Kings, Alex Thornton, Joah Madden, Applications of machine learning in animal behaviour studies, Animal Behaviour 124 (2017) 203–220.

[68] Dimitrios Kateris, Dimitrios Moshou, Xanthoula-Eirini Pantazi, Ioannis Gravalos, Nader Sawalhi, Spiros Loutridis, A machine learning approach for the condition monitoring of rotating machinery, Journal of Mechanical Science and Technology 28 (1) (2014) 61–71.

[69] Seunghee Park, Jong-Jae Lee, Chung-Bang Yun, Daniel J. Inman, A built-in active sensing system-based structural health monitoring technique using statistical pattern recognition, Journal of Mechanical Science and Technology 21 (6) (2007) 896–902.

[70] Hua Wang, Cuiqin Ma, Lijuan Zhou, A brief review of machine learning and its application, in: 2009 International Conference on Information Engineering and Computer Science, IEEE, 2009, pp. 1–4.

[71] Abu S. Islam, Kevin C. Craig, Damage detection in composite structures using piezoelectric materials (and neural net), Smart Materials and Structures 3 (3) (1994) 318.

[72] V. Cherkaasky, F. Mulier, Learning from Data, Wiley, NY, 1998.

[73] J. Ghaboussi, J.H. Garrett Jr., Xiping Wu, Knowledge-based modeling of material behavior with neural networks, Journal of Engineering Mechanics 117 (1) (1991) 132–153.

[74] Yean-Ren Hwang, Kuo-Kuang Jen, Yu-Ta Shen, Application of cepstrum and neural network to bearing fault detection, Journal of Mechanical Science and Technology 23 (10) (2009) 2730.

[75] Hua Su, Kil To Chong, Neural network based expert system for induction motor faults detection, Journal of Mechanical Science and Technology 20 (7) (2006) 929–940.

[76] Kil To Chong, Alexander G. Parlos, Comparison of traditional and neural network approaches to stochastic nonlinear system identification, KSME International Journal 11 (3) (1997) 267–278.

[77] Hui Li, Yuping Zhang, Haiqi Zheng, Gear fault detection and diagnosis under speed-up condition based on order cepstrum and radial basis function neural network, Journal of Mechanical Science and Technology 23 (10) (2009) 2780–2789.

[78] Soo-Young Kim, Byung-Young Moon, Duk-Eun Kim, Optimum design of ship design system using neural network method in initial design of hull plate, KSME International Journal 18 (11) (2004) 1923–1931.

[79] A. Chukwujekwu Okafor, K. Chandrashekhara, Y.P. Jiang, Delamination prediction in composite beams with built-in piezoelectric devices using modal analysis and neural network, Smart Materials and Structures 5 (3) (1996) 338.

[80] Farzad Pashmforoush, Ramin Khamedi, Mohamad Fotouhi, Milad Hajikhani, Mehdi Ahmadi, Damage classification of sandwich composites using acoustic emission technique and k-means genetic algorithm, Journal of Nondestructive Evaluation 33 (4) (2014) 481–492.

[81] Daniel Sammons, William P. Winfree, Eric Burke, Shuiwang Ji, Segmenting delaminations in carbon fiber reinforced polymer composite ct using convolutional neural networks, in: AIP Conference Proceedings, vol. 1706, AIP Publishing LLC, 2016, 110014.

[82] Mohamad Fotouhi, Hossein Heidary, Mehdi Ahmadi, Farzad Pashmforoush, Characterization of composite materials damage under quasi-static three-point bending test using wavelet and fuzzy c-means clustering, Journal of Composite Materials 46 (15) (2012) 1795–1808.

[83] H.N. Bar, M.R. Bhat, C.R.L. Murthy, Identification of failure modes in GFRP using PVDF sensors: Ann approach, Composite Structures 65 (2) (2004) 231–237.

PART 2

Energy harvesting

CHAPTER 3

Energy harvesting

Contents

3.1.	Introduction	41
3.2.	Sources for energy harvesting	43
	3.2.1 Piezoelectric	44
	3.2.2 Kinetic	44
	3.2.3 Photovoltaics	46
	3.2.4 Thermoelectrics	46
	3.2.5 Antennas	46
	3.2.6 Vibrations	46
	3.2.7 Wind	46
3.3.	Mechanical energy harvesting	46
	3.3.1 Mechanical modulation principle	47
3.4.	Fluid–structure interaction	48
3.5.	Thermal energy	49
3.6.	Photovoltaic technology	50
3.7.	Acoustic energy	51
3.8.	Radio frequency energy	53
3.9.	Security threats for energy harvesting system	54
References		55

3.1 Introduction

The world demand for energy is greatly increasing due to population growth and economic development. It should be noted that in one century the world has grown by 2 billion and developing countries have contributed a great deal. Preventing an oil crisis is one of the 21st century's most critical topics. Energy demand is therefore rising rapidly in line with growing population needs worldwide. Different nations around the world have their own global agendas, plans, legislation, and regulation mechanisms. The services available worldwide are diminishing as a result of demographic growth and development initiatives [1,2].

Therefore, it is very important to consider energy sources as they play a key role in meeting the need of the earth and the population. For many factors, such as the developmental profile of a nation, people's economic position, and the essence of the country's technical progress, accessible energy is not adequate for people. The release of diverse fossil fuel gases,

Piezoelectric Aeroelastic Energy Harvesting
https://doi.org/10.1016/B978-0-12-823968-1.00014-3

which can be readily available and are widely used to meet energy demand worldwide, pollutes the ecosystem significantly [3]. Developing nations are also being pressured to look for energy sources because their population is growing and they are trying to achieve economic prosperity [4].

When industrial expansion takes place, since it is proportional to economic growth, energy demand also rises. While many strategies to increase electricity generating capacity are being proposed, many people are still living in non-electrified places. The introduction of non-renewable sources of energy would certainly not satisfy energy demand as they are comprehensive and offer minimal energy supply [5].

The countries should be able to use their wealth to meet their energy demand in order to provide a long-term climate conducive to human survival. However, this is not being adequately performed at the moment since many countries depend on electricity supplies as comprehensive as clean energy sources. It is well established that a number of complicated problems, leading to a serious catastrophe, occur between countries when dominant parties prefer to enter places rich in fossil fuel reserves. In addition, the use of non-renewable power sources in containment could contribute to climate change and, in turn, serious natural disasters that damage the planet's ecosystems [6]. Therefore, it is crucial to use environmentally sustainable energy sources to improve the climate of the future [7]. In this respect, it is of crucial importance to look at alternative energies such as solar, wind, and hydroelectric power [8].

Energy harvesting is one of the most critical phenomena that researchers are working on in different fields of interest, ranging from civil to aerospace engineering. Thanks to micro- and nano-electronic devices, the consumption of electricity is reduced and now these devices can be operated directly by utilizing various energy harvesting mechanisms, i.e., piezoelectric energy harvesting, or batteries, i.e., supercapacitors. Common energy sources are represented in Fig. 3.1, and the distribution of renewable energy sources is represented in Fig. 3.2 [1]. Energy harvesting sources are presented in Fig. 3.3.

Low energy harvesting interests currently apply to independent sensor networks. In such applications, the energy harvesting scheme carries the energy captured in the capacitor and then boosted/regulated to the second storage capacitor or battery for use in the microprocessor or data transmission. Power is usually used in a sensor application and the data stored or possibly transmitted by a wireless method. The mechanisms for low power energy harvesting are presented in Fig. 3.4.

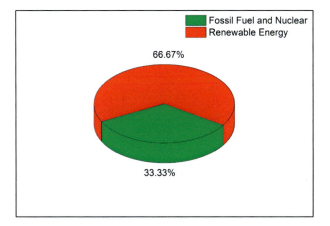

Figure 3.1 Energy sources.

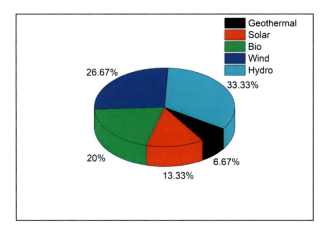

Figure 3.2 Renewable sources of energy.

3.2 Sources for energy harvesting

There are several small-scale energy sources which, in general, cannot be scaled up to industrial size in terms of output such as solar, wind, or wave power. The energy harvested by the major industrial generators can be considered infinite with respect to the low energy harvesters. Some of the common sources of energy are described in this section. These sources of energy harvesting are represented in Fig. 3.5.

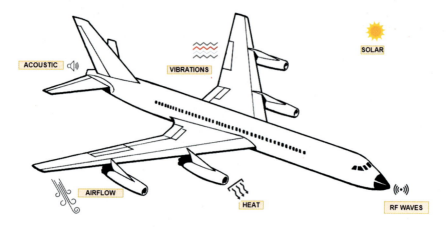

Figure 3.3 Renewable sources of energy harvesting in aerospace industry.

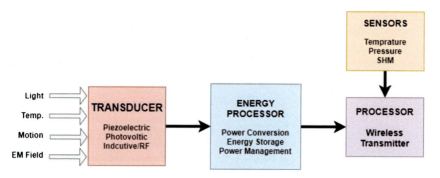

Figure 3.4 Mechanisms of energy harvesting: light, temperature, motion, or electromagnetic (EM) field are exposed to energy processors which convert them into useful electrical energy.

3.2.1 Piezoelectric

When these materials are mechanically distorted, they generate a small voltage. Engine vibration can stimulate piezoelectric materials, such as tickle by the fingers, or the fluid flow to the structure.

3.2.2 Kinetic

The motion of the structure/part can be used as a source of energy, i.e., some wristwatches use the movement of the arm for energy harvesting to drive the wristwatch.

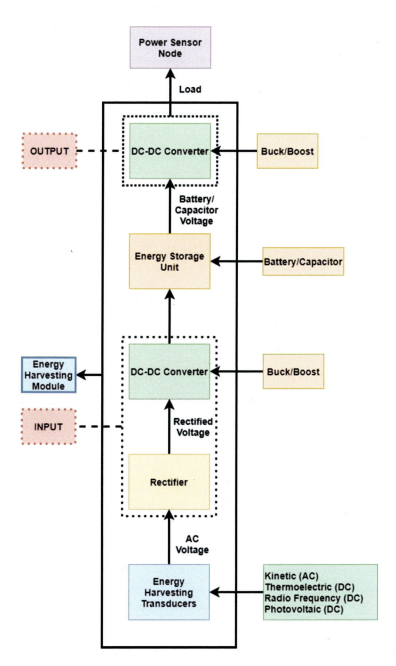

Figure 3.5 Various sources of energy harvesting.

3.2.3 Photovoltaics

Photovoltaics offer a means of harvesting electricity by transforming the use of semiconductors in solar radiation indoors/outdoors.

3.2.4 Thermoelectrics

Thermoelectrics provide a means of harvesting electrical energy by transforming the heat energy. They consist of two separate materials joined together and the thermal gradient.

3.2.5 Antennas

Energy harvesting can also be achieved with a Rectenna, and in principle with even higher frequency EM radiation by a Nantenna; special antennas can capture energy from stray radio waves.

3.2.6 Vibrations

Vibration energy obtained using electromagnetic induction in the most basic versions uses a copper coil and a magnet to produce a current which can be transformed into electricity.

3.2.7 Wind

In order to capture wind power readily accessible in the environment, wind micro-turbines are used as kinetic energy harvesters to power low-performance electronic equipment, i.e., wireless sensor nodes.

3.3 Mechanical energy harvesting

Mechanical energy harvesting in industry and academia is now widely attractive. The low-power output, low efficiency, poor environmental adaptability, and low reliability are the key problems that limit their practical applications. Mechanical energy is generated by the ambient atmosphere in fluids, vibrations, and motion. In general, they are not ideal for direct electricity conversion because the excitation may be too poor and the transducer cannot operate efficiently, or the frequency of excitation is far from the transducer resonance, or because the transducer is subjected to a high impact. Thus the mechanical energy from the environment is often processed properly in the mechanical domain, i.e., using mechanical modulation. Typical electromechanical energy transducers then transform it into

Energy harvesting 47

Figure 3.6 Various sources of mechanical energy harvesting.

electric energy [9]. The different mechanical energy harvesting sources are represented in Fig. 3.6.

3.3.1 Mechanical modulation principle

Fig. 3.6 represents the various sources of energy harvesting, i.e., wind energy, human walking, suspension systems, sensor networks, wearable, and roads [10–15]. Any of these sources of energy are not convenient to be used for direct electromechanical transfer. The human and wind-generated mechanical movement/vibration is usually small and low, so the excitation force and frequency will need to be increased with a mechanical device to maximize the output power. Broad reciprocal movement, such as vehicle suspension of cars, maybe mechanically transferred to other modes of movement, not only increasing energy densities but also optimizing design stability for inclusion in the original device for energy harvesting.

The system is sufficiently resilient to resist bad operating conditions, and complex excitation is mechanically transformed to strengthen regulation of the excitation force and minimize damage to vital components in order to increase the robustness of the energy harvesting system [16–18].

48 Piezoelectric Aeroelastic Energy Harvesting

For special methods of electromechanical transfer, which can be called mechanical modulation, including excitation form conversion, frequency up-conversion, strength/motion amplification, mechanical energy can be harvested more quickly and more effectively in a surrounding environment.

The conversion of the excitation form may also increase the frequency or power. In this way, the effectiveness of energy harvesters in applications, such as blue energy [19–21], wireless sensor networks [22,23], aerospace [24–27], biomedical [28], smart devices for aeroelasticity [29–35], smart cities [36], and SHM [37] are encouraged in mechanical modulation. The efficiency of the system for mechanical modulation can be defined as

$$\eta_{sys} = \eta_{mech}\eta_{elec}\eta_{trans}, \tag{3.1}$$

where η_{sys} is the overall efficiency of the system, η_{mech} is the mechanical modulation efficiency, and η_{trans} is the efficiency of the transducer.

3.4 Fluid–structure interaction

Energy from a fluid flow can be extracted by means of different mechanisms in order to produce electric energy. In particular, among those mechanisms we can find electromagnetic induction and piezoelectric transduction.

Typically, it entails transforming fluid dynamics actions to a spinning motion or vibration. From a very high level point of view fluid energy harvesting systems provide an output capacity that is connected to the flux speed, mechanical structure, and mechanical conversion by the following relation:

$$P \propto U^3 K_M K_{EM}, \tag{3.2}$$

where P is the output power of the fluid–structure interaction system, U^1 is the fluid flow, while K_M and K_{EM} are respectively the mechanical and electromechanical coupling coefficients of the transducer. The mechanical conversion structure of fluid energy processing has a major impact due to the complexity of the fluid–structure coupling [38]. Several piezoelectrical energy harvesters, such as flutter [39], vortex-induced vibrations [40], galloping [41,42], and wake galloping [43], were built for extracting energy from flow-induced vibrations. These mechanisms are expressed in Fig. 3.7 and are further elaborated in the later chapters.

---

[1] Observe that in Eq. (3.2) the flow velocity appears under a power of three: this is due to the fact that the flud dynamics forcers depend on U^2.

Energy harvesting 49

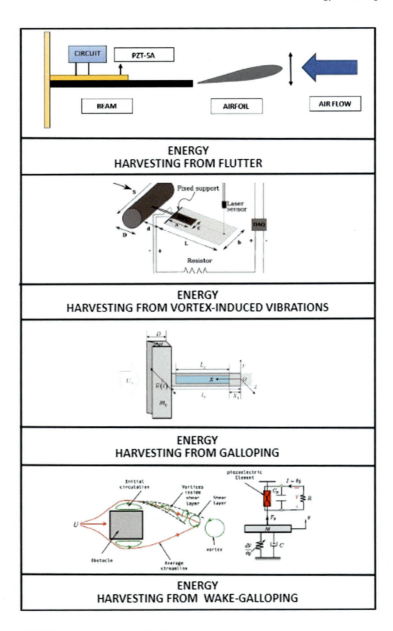

Figure 3.7 Various mechanisms for fluid–structure interaction.

3.5 Thermal energy

The piezoelectric materials, i.e., PZT-5A, BaTiO$_3$, and PVDF, showed a response that their energy harvesting capability can be increased or decreased by variation in the thermal parameters [44]. The pyroelectric coefficient λ can be used to determine the source current I_p as represented in Eq. (3.3), where T is the temperature and S is the cell's electrode surface,

$$I_p = S\lambda \frac{dT}{dt}. \tag{3.3}$$

Leonov and Fiorini have developed equations to balance thermoelectric energy collectors in analogy to electrical matching [45]. Thermal matching is necessary to optimize the extracted power and is used in the design/development of the thermal generators (TEGs) [46]. Jovanovic et al. introduced a low-power thermoelectric generator prototype. It is completely compliant with conventional microelectronic technology with a 10 μW power output [47].

3.6 Photovoltaic technology

This is a technique that converts sunlight directly into electrical energy without a conversion interface. So, these instruments are very simple to manufacture and handle. Moreover, they can produce large output from relatively small input. They are therefore used globally in a number of applications [48].

Photovoltaic systems typically use semiconductor materials to induce electricity, and commonly use silicon. The theory of this system is that electrons are enabled by providing extra energy. The system operates under the theory that electrons are stimulated by the energy supply from sunlight, from a lesser energy state to a higher energy state. In exchange, this activation would cause the semiconductor to provide electricity with numbers of holes and electrons [49].

The selection of the materials for the semiconductors is based on many parameters, factors, and conditions [50]. The photovoltaic (PV) system consists of a wide variety of elements such as batteries, modules, and power generation arrays. Various methods of regulation and monitoring are often used to enhance operating performance by means of electronic instruments, electrical contacts, and mechanical equipment. PV systems are measured in peak kW, which is a volume of electric power transmitted by a PV device while the sun is overhead directly [51]. A lot of research has been conducted on photovoltaic devices during the last several years to improve the performance of these devices, which is now reportedly a rapidly expanding industry with an average growth of 48% in demand since 2002 [52,53].

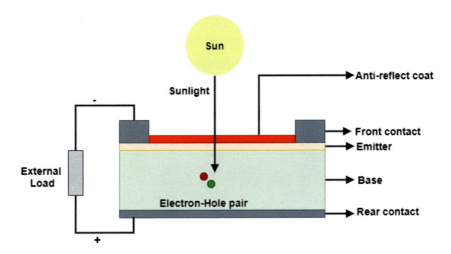

Figure 3.8 Solar cell operation.

Besides PV, solar thermal and photovoltaic concentrations are applied to convert solar energy into electricity [54]. The overall representation of a working solar cell is provided in Fig. 3.8.

3.7 Acoustic energy

Because of their abundance and purity, sound energy sources have a huge capacity for energy harvesting. But, due to their low energy density, sound methods of energy production are less studied than the use of the other energy sources. Various methods are conducted worldwide to address the low energy density of the signal [55]. Sound technology for energy recovery is expressed graphically in Fig. 3.9. Initially, an environmental sound wave is gathered, amplified, and transformed into electric power by the resonant or acoustic metameter process. Finally, for applications involving power electronics, rectification, control, and energy conservation are carried out because energy is present in an alternating current (AC).

As energy demand is being lowered, like in wireless sensor networks, and therefore the efficiency of energy harvesting can be improved by the proper design of the harvester, sound energy collection has a great potential for use in some applications like powering low-power machines, and it can solve the noise issue as well. Audible frequency spectrum (20–20,000 Hz) and low-frequency sounds (< 500 Hz) are helpful in sound energy harvesting because the low-frequency sound is predominantly noisy and

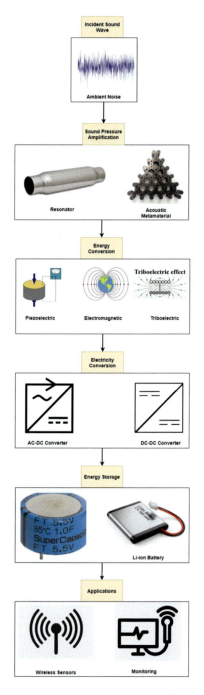

Figure 3.9 Schematic diagram for overall acoustic energy harvesting.

difficult to suppress, so that it is well-transmitted [56,57]. The degree of the sound pressure is a logarithmic manifestation of the sound pressure relative to a reference value and defined by

$$P_{SL} = 10\log_{10}(\frac{p}{p_o})^2 = 20\log_{10}\frac{p}{p_o}, \qquad (3.4)$$

where P_{SL} is the sound pressure level, p is the root mean square sound pressure and p_o has the reference value of 20 μPa.

The resonator, membrane, and piezoelectric materials are typically used in sound energy harvesters. At its resonant frequency, the resonator induces sound oscillation. A membrane oscillates with sound oscillation, on the one hand, and with a piezoelectric material, on the other. Mechanical energy is converted into electrical energy by the piezoelectric material. The received sound intensity is normally amplified by its low density with the resonators in sound energy harvesters. There are several resonator forms, like Helmholtz, VW, and HW resonator [58].

3.8 Radio frequency energy

As methods for battery-free renewable wireless networks, collecting energy from radio frequency (RFEH) and wireless power transfer (WPT) have been of major interest. The correction antennas are at the heart of WPT and RFEH systems and have a crucial impact on DC load power supply. The antenna portion of rectenna has direct effects on the efficiency of radio frequency harvesting, which can differ by an order of magnitude in harvesting capacity [59]. The source (which can be an electrical unit or a circuit as represented in Fig. 3.10) transmits RF signals and the implementation circuit that has an integrated circuit for the transfer of energy receives an RF which then makes a possible difference through the antenna length and produces a charge carrier movement through the antenna. The charger transmits the RF to DC, i.e., the charge is now converted to DC current by the circuit and is temporarily saved in the condenser. Then the energy is amplified or converted to the potential value as requested by the load through the energy conditioning circuit.

Many channels relay RF signals, such as satellite, radio, wireless Internet. Any application with the RF power harvesting circuit will receive the signal and transform it into electricity. The conversion phase starts when the transmitting antenna collects the signal and causes a possible change

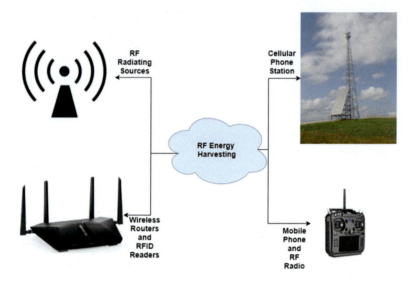

Figure 3.10 Sources for RF energy harvesting.

over the antenna length, which then further lets the antenna load carriers shift. These antenna charging carriers switch to the circuit connected to the impedance by the cables. Power transmission from an antenna (RF source) through a corrector/spanner multiplier (load) shall be maximized by the impedance matching network (IMN). For optimal power transfers between source and load, the impedance in an RF circuit is as significant as the resistance in the DC circuit.

A sinusoidal waveform RF signal is obtained on the antenna, i.e., an AC signal which must be turned into a DC signal. Once the rectifier/voltage multiplier circuit is passed into IMN, the signal is rectified and amplified when required. The corrective circuit is a voltage multiplier circuit (a special corrective) that corrects and also boosts the corrected signal depending on the application requirement. The rectifier circuit is not a half-wave, full-wave, or bridge rectifier. The electricity converted from AC to DC with a voltage converter is transferred to the circuit for power storage that uses a capacitor or battery to store and supply electricity to charge (application) whenever necessary.

3.9 Security threats for energy harvesting system

The assaults on various protocol stack layers are revealed to wireless networks and computers using energy harvesting. These attacks go way beyond

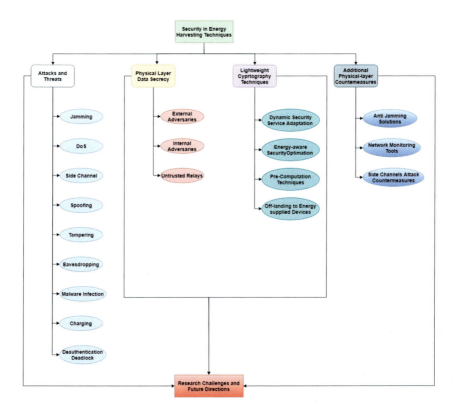

Figure 3.11 Security threats for energy harvesting system.

reliability and can jeopardize the overall accessibility, reliability, and reliability of the network [16]. Security threats for energy harvesting systems are presented in Fig. 3.11.

References

[1] Nadarajah Kannan, Divagar Vakeesan, Solar energy for future world: a review, Renewable and Sustainable Energy Reviews 62 (2016) 1092–1105.
[2] Shahriar Shafiee, Erkan Topal, When will fossil fuel reserves be diminished?, Energy Policy 37 (1) (2009) 181–189.
[3] Martin M. Halmann, Meyer Steinberg, Greenhouse Gas Carbon Dioxide Mitigation: Science and Technology, CRC Press, 1998.
[4] John Asafu-Adjaye, The relationship between energy consumption, energy prices and economic growth: time series evidence from Asian developing countries, Energy Economics 22 (6) (2000) 615–625.
[5] A.B. Etemoglu, Thermodynamic evaluation of geothermal power generation systems in turkey, Energy Sources, Part A 30 (10) (2008) 905–916.

[6] Poul Schou, Polluting non-renewable resources and growth, Environmental and Resource Economics 16 (2) (2000) 211–227.

[7] Kari Alanne, Arto Saari, Distributed energy generation and sustainable development, Renewable and Sustainable Energy Reviews 10 (6) (2006) 539–558.

[8] Antonia V. Herzog, Timothy E. Lipman, Daniel M. Kammen, et al., Renewable energy sources, in: Encyclopedia of Life Support Systems (EOLSS). Forerunner Volume – Perspectives and Overview of Life Support Systems and Sustainable Development, 2001, p. 76.

[9] Hong-Xiang Zou, Lin-Chuan Zhao, Qiu-Hua Gao, Lei Zuo, Feng-Rui Liu, Ting Tan, Ke-Xiang Wei, Wen-Ming Zhang, Mechanical modulations for enhancing energy harvesting: principles, methods and applications, Applied Energy 255 (2019) 113871.

[10] Mohsin Ali Marwat, Weigang Ma, Pengyuan Fan, Hassan Elahi, Chanatip Samart, Bo Nan, Hua Tan, David Salamon, Baohua Ye, Haibo Zhang, Ultrahigh energy density and thermal stability in sandwich-structured nanocomposites with dopamine@Ag@BaTiO$_3$, Energy Storage Materials 31 (2020) 492–504.

[11] Ahsan Ali, Riffat Asim Pasha, Hassan Elahi, Muhammad Abdullah Sheeraz, Saima Bibi, Zain Ul Hassan, Marco Eugeni, Paolo Gaudenzi, Investigation of deformation in bimorph piezoelectric actuator: analytical, numerical and experimental approach, Integrated Ferroelectrics 201 (1) (2019) 94–109.

[12] Muhammad Usman Khan, Zubair Butt, Hassan Elahi, Waqas Asghar, Zulkarnain Abbas, Muhammad Shoaib, M. Anser Bashir, Deflection of coupled elasticity–electrostatic bimorph PVDF material: theoretical, FEM and experimental verification, Microsystem Technologies 25 (8) (2019) 3235–3242.

[13] Vittorio Memmolo, Hassan Elahi, Marco Eugeni, Ernesto Monaco, Fabrizio Ricci, Michele Pasquali, Paolo Gaudenzi, Experimental and numerical investigation of PZT response in composite structures with variable degradation levels, Journal of Materials Engineering and Performance 28 (6) (2019) 3239–3246.

[14] H. Elahi, A. Israr, R.F. Swati, H.M. Khan, A. Tamoor, Stability of piezoelectric material for suspension applications, in: 2017 Fifth International Conference on Aerospace Science & Engineering (ICASE), IEEE, 2017, pp. 1–5.

[15] Hassan Elahi, Marco Eugeni, Paolo Gaudenzi, Electromechanical degradation of piezoelectric patches, in: Analysis and modelling of advanced structures and smart systems, Springer, 2018, pp. 35–44.

[16] Hassan Elahi, Khushboo Munir, Marco Eugeni, Sofiane Atek, Paolo Gaudenzi, Energy harvesting towards self-powered IoT devices, Energies 13 (21) (2020) 5528.

[17] Hassan Elahi, Marco Eugeni, Paolo Gaudenzi, A review on mechanisms for piezoelectric-based energy harvesters, Energies 11 (7) (2018) 1850.

[18] Zubair Butt, Riffat Asim Pasha, Faisal Qayyum, Zeeshan Anjum, Nasir Ahmad, Hassan Elahi, Generation of electrical energy using lead zirconate titanate (PZT-5a) piezoelectric material: analytical, numerical and experimental verifications, Journal of Mechanical Science and Technology 30 (8) (2016) 3553–3558.

[19] J. Sarmiento, A. Iturrioz, V. Ayllón, R. Guanche, I.J. Losada, Experimental modelling of a multi-use floating platform for wave and wind energy harvesting, Ocean Engineering 173 (2019) 761–773.

[20] Youn-Hwan Shin, Inki Jung, Myoung-Sub Noh, Jeong Hun Kim, Ji-Young Choi, Sangtae Kim, Chong-Yun Kang, Piezoelectric polymer-based roadway energy harvesting via displacement amplification module, Applied Energy 216 (2018) 741–750.

[21] Muhammad Umer Sohail, Hassan Elahi, Asad Islam, Hossein Raza Hamdani, Khalid Parvez, Raees Fida Swati, CFD analysis on the effects of distorted inlet flows with variable rpm on the stability of the transonic micro-compressor, Microsystem Technologies (2021) 1–17.

[22] Lifeng Wang, Kai-Kit Wong, Shi Jin, Gan Zheng, Robert W. Heath, A new look at physical layer security, caching, and wireless energy harvesting for heterogeneous ultra-dense networks, IEEE Communications Magazine 56 (6) (2018) 49–55.

[23] Canan Dagdeviren, Zhou Li, Zhong Lin Wang, Energy harvesting from the animal/human body for self-powered electronics, Annual Review of Biomedical Engineering 19 (2017) 85–108.

[24] Hassan Elahi, Khushboo Munir, Marco Eugeni, Muneeb Abrar, Asif Khan, Adeel Arshad, Paolo Gaudenzi, A review on applications of piezoelectric materials in aerospace industry, Integrated Ferroelectrics 211 (1) (2020) 25–44.

[25] Hassan Elahi, Marco Eugeni, Paolo Gaudenzi, Madiha Gul, Raees Fida Swati, Piezoelectric thermo electromechanical energy harvester for reconnaissance satellite structure, Microsystem Technologies 25 (2) (2019) 665–672.

[26] Hassan Elahi, Marco Eugeni, Paolo Gaudenzi, Faisal Qayyum, Raees Fida Swati, Hayat Muhammad Khan, Response of piezoelectric materials on thermomechanical shocking and electrical shocking for aerospace applications, Microsystem Technologies 24 (9) (2018) 3791–3798.

[27] Hassan Elahi, Zubair Butt, Marco Eugnei, Paolo Gaudenzi, Asif Israr, Effects of variable resistance on smart structures of cubic reconnaissance satellites in various thermal and frequency shocking conditions, Journal of Mechanical Science and Technology 31 (9) (2017) 4151–4157.

[28] Mohsen Safaei, R. Michael Meneghini, Steven R. Anton, Energy harvesting and sensing with embedded piezoelectric ceramics in knee implants, IEEE/ASME Transactions on Mechatronics 23 (2) (2018) 864–874.

[29] Hassan Elahi, Khushboo Munir, Marco Eugeni, Paolo Gaudenzi, Reliability risk analysis for the aeroelastic piezoelectric energy harvesters, Integrated Ferroelectrics 212 (1) (2020) 156–169.

[30] Hassan Elahi, Marco Eugeni, Luca Lampani, Paolo Gaudenzi, Modeling and design of a piezoelectric nonlinear aeroelastic energy harvester, Integrated Ferroelectrics 211 (1) (2020) 132–151.

[31] Hassan Elahi, Marco Eugeni, Federico Fune, Luca Lampani, Franco Mastroddi, Giovanni Paolo Romano, Paolo Gaudenzi, Performance evaluation of a piezoelectric energy harvester based on flag-flutter, Micromachines 11 (10) (2020) 933.

[32] Hassan Elahi, The investigation on structural health monitoring of aerospace structures via piezoelectric aeroelastic energy harvesting, Microsystem Technologies (2020) 1–9.

[33] Marco Eugeni, Hassan Elahi, Federico Fune, Luca Lampani, Franco Mastroddi, Giovanni Paolo Romano, Paolo Gaudenzi, Numerical and experimental investigation of piezoelectric energy harvester based on flag-flutter, Aerospace Science and Technology 97 (2020) 105634.

[34] Hassan Elahi, Marco Eugeni, Paolo Gaudenzi, Design and performance evaluation of a piezoelectric aeroelastic energy harvester based on the limit cycle oscillation phenomenon, Acta Astronautica 157 (2019) 233–240.

[35] Marco Eugeni, Hassan Elahi, Federico Fune, Luca Lampani, Franco Mastroddi, Giovanni Paolo Romano, Paolo Gaudenzi, Experimental evaluation of piezoelectric energy harvester based on flag-flutter, in: Conference of the Italian Association of Theoretical and Applied Mechanics, Springer, 2019, pp. 807–816.

[36] Mingyi Liu, Rui Lin, Shengxi Zhou, Yilun Yu, Aki Ishida, Margarita McGrath, Brook Kennedy, Muhammad Hajj, Lei Zuo, Design, simulation and experiment of a novel high efficiency energy harvesting paver, Applied Energy 212 (2018) 966–975.

[37] Muhammad Iqbal, Farid Ullah Khan, Hybrid vibration and wind energy harvesting using combined piezoelectric and electromagnetic conversion for bridge health monitoring applications, Energy Conversion and Management 172 (2018) 611–618.

[38] Abdessattar Abdelkefi, Zhimiao Yan, Muhammad R. Hajj, Modeling and nonlinear analysis of piezoelectric energy harvesting from transverse galloping, Smart Materials and Structures 22 (2) (2013) 025016.

[39] Soon-Duck Kwon, A t-shaped piezoelectric cantilever for fluid energy harvesting, Applied Physics Letters 97 (16) (2010) 164102.

[40] A. Mehmood, A. Abdelkefi, M.R. Hajj, A.H. Nayfeh, I. Akhtar, A.O. Nuhait, Piezoelectric energy harvesting from vortex-induced vibrations of circular cylinder, Journal of Sound and Vibration 332 (19) (2013) 4656–4667.

[41] A. Abdelkefi, M.R. Hajj, A.H. Nayfeh, Piezoelectric energy harvesting from transverse galloping of bluff bodies, Smart Materials and Structures 22 (1) (2012) 015014.

[42] Abdessattar Abdelkefi, Muhammad R. Hajj, Ali H. Nayfeh, Power harvesting from transverse galloping of square cylinder, Nonlinear Dynamics 70 (2) (2012) 1355–1363.

[43] Abdessattar Abdelkefi, John Michael Scanlon, E. McDowell, Muhammad R. Hajj, Performance enhancement of piezoelectric energy harvesters from wake galloping, Applied Physics Letters 103 (3) (2013) 033903.

[44] A. Cuadras, M. Gasulla, Vittorio Ferrari, Thermal energy harvesting through pyroelectricity, Sensors and Actuators A: Physical 158 (1) (2010) 132–139.

[45] V. Leonov, P. Fiorini, Thermal matching of a thermoelectric energy scavenger with the ambience, in: Proceedings of the European Conference on Thermoelectrics, 2007, pp. 129–133.

[46] Matthias Stordeur, Ingo Stark, Low power thermoelectric generator-self-sufficient energy supply for micro systems, in: XVI ICT'97, Proceedings ICT'97, 16th International Conference on Thermoelectrics (Cat. No. 97TH8291), IEEE, 1997, pp. 575–577.

[47] Velimir Jovanovic, Saeid Ghamaty, Design, fabrication, and testing of energy-harvesting thermoelectric generator, in: Smart Structures and Materials 2006: Smart Structures and Integrated Systems, vol. 6173, 2006, 61730G.

[48] Philip P. Barker, James M. Bing, Advances in solar photovoltaic technology: an applications perspective, in: IEEE Power Engineering Society General Meeting, 2005, IEEE, 2005, pp. 1955–1960.

[49] Xiuling Li, Metal assisted chemical etching for high aspect ratio nanostructures: a review of characteristics and applications in photovoltaics, Current Opinion in Solid State and Materials Science 16 (2) (2012) 71–81.

[50] Takhir M. Razykov, Chris S. Ferekides, Don Morel, Elias Stefanakos, Harin S. Ullal, Hari M. Upadhyaya, Solar photovoltaic electricity: current status and future prospects, Solar Energy 85 (8) (2011) 1580–1608.

[51] Bhubaneswari Parida, Selvarasan Iniyan, Ranko Goic, A review of solar photovoltaic technologies, Renewable and Sustainable Energy Reviews 15 (3) (2011) 1625–1636.

[52] Robert Kropp, Solar expected to maintain its status as the world's fastest-growing energy technology, Sustainability Investment News (2009).

[53] Hassan Elahi, Ali Tamoor, Abdul Basit, Asif Israr, Raess Fida Swati, Shamraiz Ahmad, Usman Ghafoor, Muhammad Shaban, Design and performance analysis of hybrid solar powered geyser in Islamabad, Pakistan, Thermal Science 00 (2018) 299.

[54] Xinping Zhou, Yangyang Xu, Solar updraft tower power generation, Solar Energy 128 (2016) 95–125.

[55] Jaehoon Choi, Inki Jung, Chong-Yun Kang, A brief review of sound energy harvesting, Nano Energy 56 (2019) 169–183.

[56] Ming Yuan, Hongli Ji, Jinhao Qiu, Tianbing Ma, Active control of sound transmission through a stiffened panel using a hybrid control strategy, Journal of Intelligent Material Systems and Structures 23 (7) (2012) 791–803.

[57] Liang Fang, Jian Liu, Sheng Ju, Fengang Zheng, Wen Dong, Mingrong Shen, Experimental and theoretical evidence of enhanced ferromagnetism in sonochemical synthesized $BiFeO_3$ nanoparticles, Applied Physics Letters 97 (24) (2010) 242501.

[58] Aichao Yang, Ping Li, Yumei Wen, Caijiang Lu, Xiao Peng, Wei He, Jitao Zhang, Decai Wang, Feng Yang, Note: High-efficiency broadband acoustic energy harvesting using Helmholtz resonator and dual piezoelectric cantilever beams, Review of Scientific Instruments 85 (6) (2014) 066103.

[59] Adamu Murtala Zungeru, Li-Minn Ang, S. Prabaharan, Kah Phooi Seng, Radio frequency energy harvesting and management for wireless sensor networks, in: Green Mobile Devices and Networks: Energy Optimization and Scavenging Techniques, vol. 13, CRC Press, New York, NY, USA, 2012, pp. 341–368.

CHAPTER 4

Piezoelectric energy harvesters

Contents

4.1.	Introduction	61
4.2.	Piezoceramics-based energy harvesting	62
	4.2.1 Cymbal type	64
	4.2.2 Cantilever-type vibration energy harvesting	65
	4.2.3 Modeling and theory	65
	4.2.4 New materials for energy harvesting	67
4.3.	Energy harvesting with piezopolymers	68
4.4.	Harvesting model	70
	4.4.1 Governing equations	70
References		74

4.1 Introduction

Energy harvesting is described as the collection, accumulation, and storage of minute quantities of energy from single or multiple surrounding energy sources. Power harvesting and energy scavenging are other terms for energy harvesting. With recent developments in wireless and MEMS technologies, energy storage is being introduced as a feasible alternative to the standard battery. The wireless sensors and portable electronics devices with ultra-low power use conventional batteries as their source of power, but their life is small in comparison to the operating life of the computers. It is inefficient and at times impossible to repair or recharge the battery. Therefore, a large number of research projects have been implemented into the technologies of harvesting energy as a self-powered source for systems of wireless sensor networks or portable devices.

With respect to energy conversion, people have already made use of windmill, solar energy, geothermal, and watermill technologies for harvesting energy. As PZT material has the ability to transform a mechanical vibration into electricity with simple structures, it is noted that PZT energy can be used as a self-powered source for the wireless network sensors [1–6]. Piezoelectricity is basically the pressure electricity and also represents crystalline properties of materials such as tourmaline, quartz, barium titanate, and Rochelle salt which, because of applied pressure, develop electricity [7,8]. The phenomenon is called the direct effect, whereas if an electric

Piezoelectric Aeroelastic Energy Harvesting
https://doi.org/10.1016/B978-0-12-823968-1.00015-5

Copyright © 2022 Elsevier Inc.
All rights reserved.

field is applied then there can be some deformation in these crystals known as the converse effect. The direct effect is being used in a transducer or sensor, and the converse effect is utilized in the actuator. Two linearized constitutive equations can be used to model the piezoelectric materials coupled with electromechanical behavior [7] in which the second, third and fourth order tensorial components are arranged in matrix notation:

$$\underline{D} = \underline{\underline{e}}^{\sigma} \underline{E} + \underline{\underline{d}} \, \underline{\sigma}. \tag{4.1}$$

Eq. (4.1) represents the direct piezoelectric effect. One also has

$$\underline{\epsilon} = \underline{\underline{d}} \, \underline{E} + \underline{\underline{S}}^{E} \, \underline{\sigma}, \tag{4.2}$$

which describes the converse piezoelectric effect. Whereas dielectric displacement is represented by the vector $\underline{D}$ in C/m^2, strain in N/mV is denoted by vector $\underline{\epsilon}$, the electric field applied is represented by the vector $\underline{E}$ in V/m, stress is given by $\underline{\sigma}$ in N/m^2; $\underline{d}$ in either m/V or C/N are PZT coefficients, while the elastic compliance matrix $\underline{\underline{S}}^{E}$ in m^2/N and the dielectric permittivity $\underline{\epsilon}$ in F/m or N/V^2 are considered as the PZT constants. They have numerous applications in the field of aerospace industry [9–12] and energy harvesting [13–16].

There are two categories of piezoelectric materials, called piezopolymers and piezoceramics. Piezoceramics possesses a high coupling of electromechanical constants and gives a high rate of conversion of energy, these are porous to be used as a general form energy transducer. On the contrary, the electromechanical coupling constants for the piezopolymers are small in comparison to piezoceramics, but piezopolymers are flexible. Some material characteristics of piezopolymer (polyvinylidene fluoride, PVDF) and piezoceramics (PZT-5H, PZT-8) are given in Table 4.1.

4.2 Piezoceramics-based energy harvesting

Priya et al. [17] reviewed the detailed PZT energy harvesting coverage with low-profile transducer and findings of different prototypes for energy harvesting. The authors also provided a discussion on PZT material selection for the applications of on-and-off resonance. This theoretical calculation explains that for PZT energy harvesting devices the energy density is 3 to 5 times greater than for electromagnetic and electrostatic devices.

Wright et al. showed a variety of devices for harvesting energy from vibrations [18–21]. They began with an indication of potential power sources

Table 4.1 PZT characteristics [7].

Coefficient	PVDF	PZT-8	PZT-5H
d_{15}	–	330	741×10^{-12} m/V
d_{32}	2.5–3	−97	-274×10^{-12} m/V
d_{31}	18–24	−97	-274×10^{-12} m/V
d_{33}	−33	225	593×10^{-12} m/V
Curie temperature	–	300	193°C
Poling field DC	–	5.5	12 kV/cm
Density	–	7600	7500 kg/m^3
Free-strain range	–	$\mu\epsilon$	−250 to +850
Relative permittivity e_{33}	–	1000	3400
Depoling field AC	–	15	7 kV/cm
E_{11}	–	87	62 GPa
E_{33}	–	74	48 GPa
Tensile strength (dynamic)	–	75.8	27.6 MPa
Tensile strength (static)	–	75.8	75.8
Compressive strength (static)	–	> 517	> 517 MPa
Compressive depoling limit	–	150	> 30 MPa

from low-level vibrations in households and offices. After that they investigated piezoelectric converters and capacitive MEMS [18]. The outcome shows that harnessed power is substantially higher with piezoelectric conversion. A two-layered cantilever PZT generator was optimized by them, which was later validated using theoretical analysis [19]. Devices based on the small cantilever were modeled using PZT materials that can scavenge power from the sources with low-level ambient vibrations, and a new configuration design was presented for the increment in the capacity of energy harvesting [20]. Up to 24% of resonance frequency was decreased by using axially compressed PZT bimorph. The observed output power was 65% to 90% of nominal frequency at a frequency below the unloaded resonance frequency by 19–24% [21]. They have wide applications in the field of aeroelastic piezoelectric energy harvesting [22–28].

Monolithic PZT and microfiber composite (MFC) were investigated by Sodano et al. [29], and for both materials they estimated their efficiency. The authors did an experimental investigation for three PZT devices, i.e., MFC, bimorph QP, and monolithic PZT, to determine their recharge capacity of the discharged battery [30].

For application with low-frequency energy harvesting, Shen et al. [31] developed a PZT cantilever with the help of micromachined Si proof mass.

The obtained power density was 416 W/cm^3 and the average power was 0.32 W. The power generator array was developed by Liu et al. [32] which was based on the thick-film PZT cantilevers for the improvement of output power and flexibility of frequency. The performance of effective electric power was improved to 3.98 mW and the output voltage of 3.93 V DC to load resistance. Self-supportive sensors were enabled with the help of an energy harvesting MEMS device by Choi et al. [33] using thin-film PZT. Resonating with particular frequencies of an external source of vibrating energy will generate electrical energy through the piezoelectric effect.

4.2.1 Cymbal type

Under a transverse external force, in-plane large strain can be produced by the cymbal structure which is beneficial for micro-energy harvesting. According to Kim et al. [34], under pre-stress cyclic conditions, promising results were reported by a PZT energy harvester and experimental results were validated by finite element analysis. Two ring-type PZT stacks were presented by Li et al. [35], with one shaft for the purpose of pre-compressing and a pair of bow-shaped elastic plates. The study indicates that the piezo-electric flex-compressing mode is capable of generating a higher electrical voltage and power output than the conventional flex-tensional mode. Conventional PZT energy harvesters are represented in Figs. 4.1 and 4.2.

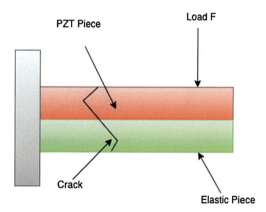

Figure 4.1 Conventional PZT energy harvesters.

4.2.2 Cantilever-type vibration energy harvesting

This type of energy harvesting is composed of a simple structure and, under vibration, can undergo a large deformation. Fundamental limitations were imposed by Flynn et al. [36] on the PZT (lead zirconate titanate), and they explained that in typical PZT materials, an effective constraint is a limit on mechanical stress.

A beam element was used by Elvin et al. [37] for theoretical modeling and experimentation of power harvesting from a PZT material. As per their results, simple beam bending is enough to provide a self-power source for the strain energy sensor.

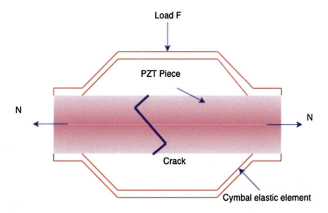

Figure 4.2 Conventional PZT energy harvesters.

4.2.3 Modeling and theory

The PZT coupling effect, structure, and electrical behavior are included in the theoretical modeling of PZT energy harvesting. A better version of the mathematical model was proposed by Erturk et al. [38], and they attempted to solve the problems related to the oversimplified formulation which is based on mathematics, for example, physical and base motion modeling, PZT coupling, and low fidelity models. Correction factors were also suggested by them for the single DoF of the model with base excitation, and the authors investigated the single DoF harmonic base excitation connections, hence widely used by researchers for extracting energy via mechanical vibration [39]. The analytical approach of a closed-form was mentioned also for piezoceramic layers connected in parallel and in series [40].

To estimate the electrical power output of plates of the PZT energy harvester, Marqui et al. [41] proposed a model with plates of finite element (FE) which is a couple electromechanically depending upon assumptions of a Kirchhoff plate which also accounts for the conductive electrodes effect. Analytical and experimental solutions for a unimorph cantilever beam are used to verify the FEA simulation findings. Renno et al. [42] improved the energy harvester based on PZT vibration by adding a resistive and an inductor load, concluding that the presence of an inductor in the circuit increases harvesting power performance.

Pouline et al. [43] related an electromagnetic device consisting of a magnet moving inside a coil to a piezoelectric system consisting of a PZT ceramic bar which at one end is free whereas on the other is constrained. A strong similarity was predicted along with a single-level duality. Ajitsaria et al. [44] predicted an analytical method for voltage and power generation depending upon the beam theory of Euler–Bernoulli and beam equations of Timoshenko. They demonstrated that a comparison of experimental and simulation outcomes was satisfactory.

By characterizing the relationship between the storage circuit and harvesting structure using a non-linear rectifier, Hu et al. [45] modeled a PZT harvester as an integrated mechanical device. They demonstrated how to optimize power density by adjusting inductance that is dimensionless with a given corresponding aspect ratio and a fixed dimensionless end mass.

Shu et al. [46] measured the energy conversion efficiency under steady-state conditions for a rectified piezoelectric power harvester. They discovered that the optimization parameters are influenced by electromechanical coupling's relative strength.

Marzencki et al. [47] suggested using mechanical nonlinear strain stiffening to create a passive, wideband adaptive device. They reported experimentally confirmed adaptability of frequency of more than 36% for a clamped–clamped beam system at $2g$ acceleration of input. According to them, the proposed solution was ideal for the industry of autonomous machinery monitoring systems, where there is an abundance of high amplitude vibrations.

To find the parameter values that are over-predicted in models of Euler–Bernoulli beam, Dietl et al. [48] suggested a Timoshenko model of the transverse piezoelectric beam. They stated the exact expression for the power, current, voltage, and tip deflection of the piezoelectric beam. They also used heuristic optimization code to optimize the shapes of beams for power harvesting; furthermore, properties of an optimal beam were validated with experimental results [49].

Gammaitoni et al. [50] used a non-linear stochastic differential equation to model the dynamics of a piezoelectric harvesting oscillator, emphasizing the benefits of noise and non-linearity. By using the theoretical analysis of harvesting energy from vibrations, Gao et al. [51] used a PZT unimorph cantilever with non-piezoelectric and unequal piezoelectric lengths. They discovered, for a given frequency of vibration, that the highest open-circuit induced voltage occurs in the case where the non-piezoelectric-to-piezoelectric length ratio is higher than unity, whereas it reaches its maximum power when this ratio becomes unity.

Knight et al. [52] proposed a method for determining the optimum energy harvesting for cantilever beams with interdigitated piezoelectric MEMS that are unimorph. It was demonstrated that poling behavior turned out to be the most significant element in evaluating the true losses referring to non-uniform poling. A parametric analysis was performed to explore the effect of patterns of an electrode, PZT layer measurements, and dimensions of the electrode on the poling factor percentage. They suggested how to design PZT MEMS devices that are optimal for energy harvesting.

Ly et al. [53] produced a model based on the bending of PZT cantilever of the 31-effect based on the Euler–Bernoulli beam principle. The global system's equations of motion were developed using Hamilton's principle and solved with the help of a decomposition modal. A mathematical model was developed that improves the mechanical energy transformation into electrical energy by the use of the direct PZT effect. It was demonstrated that the resonant frequency's second mode had a much greater voltage and bandwidth than the first mode. Richards et al. [54] formed a method for estimation of various piezogenerators based on power conversion efficiency.

4.2.4 New materials for energy harvesting

Jeong et al. [55] have examined the experimentally produced piezoelectric microstructure ceramics by the casting of slurries containing the $SrTiO_3$ (STO) prototype under the stress of external mechanism. It was claimed that when high stress is imposed on the specimen, STO-added specimens have shown excellent control over the STO-free specimen.

Because of its high resulting level of voltage, Elfrink et al. [56] researched aluminum nitride (AlN) as a PZT material for PZT energy harvesting. They registered a maximum peak power of 60 W for a device that was unpackaged at a 2.0 g acceleration and a 572 Hz resonance frequency.

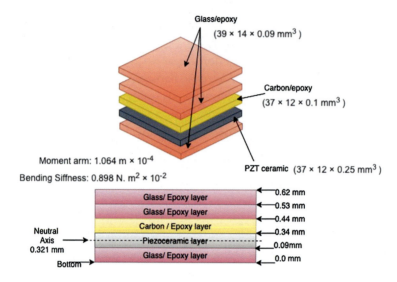

Figure 4.3 Geometry and position of the neutral axis of PCGE-A.

Tien and Goo [57] investigated a piezocomposite made of carbon/epoxy, PZT ceramic, and glass/epoxy layers to extract energy (Fig. 4.3). After numerical and experimental confirmation, they stated that piezocomposite has the ability to harvest energy subjected to vibration.

4.3 Energy harvesting with piezopolymers

Mateu et al. [58] investigated multiple bending beam configurations using piezofilms appropriate for shoe inserts and excitations based on the walking-type and calculated for each type the resulting strain as a function of geometrical parameters and material properties. The optimal configuration was found by analyzing the energy extracted. Based on their previous work, the authors created piezoelectric film inserts within a shoe [59].

Farinholt et al. [60] created a novel energy harvesting backpack that uses PVDF to produce electrical energy from the wearer's differential forces with the backpack. They also recommended a comparison of energy harvesting of PVDF with anionic polymer ionic transducer for the performance of properties of electromechanical conversion [61]. Analytical models for each material based on a spring–mass–damper were created, then experimental results were compared with simulation results.

Shah et al. [62] investigated the micropower produced by harvesting generators made of PP (polypropylene) foam polymer, PVDF (polyvinylidene fluoride) membrane, and piezoelectric ceramic (PZT). They demonstrated that PZT polymer materials have the ability to generate higher voltage/power than ceramic-based PZT materials and that PZT polymer materials can be used to generate electricity.

Sohn et al. [63] used FEM to test the harvesting power ability of piezofilm under blood pressure operation and theoretically analyzed square and circular configurations. Hu et al. [64] suggested a corrugated PVDF bimorph power harvester with an arrangement set for harvesting, at two edges in the corrugation path and other edges being free. They changed the resonant frequency by adjusting the geometrical configuration to maintain the harvester running at an optimum state. They claimed that by developing the harvesting structure with adjustable resonance, the harvester's adaptability to the operating system was improved significantly.

Liu [65] showed a harvesting system that used switch-mode power electronics to regulate PZT device charge in relation to the mechanical input for efficient conversion of energy. In such experiments, the method towards active harvesting energy is improved to obtain energy by a factor of 5 compared with the optimized diode rectifier circuit with the same mechanical displacement.

Chang et al. [66], with characteristics of electric poling and in situ mechanical stretches, generated piezoelectrical properties using near-field spinning for direct writing PVDF nanofibers. They found that nanogenerators demonstrated repeatable and consistent energy transfer outputs under mechanical stretching higher in order of magnitude than those made of thin PVDF films.

Hansen et al. [67] created a hybrid energy scavenging system that included a piezoelectric PVDF nanofiber generator for processing both biochemical and biomechanical energy. It was discovered that two forms of harvesting energy could operate concurrently or independently, increasing output and lifetime.

Chang [68] built and tested an energy harvester based on piezo-elastica in hard disk drives of a computer. Simulations of the finite numerical element and measurements of the laboratory revealed that the power of about 25% is consumed by the voice coil motor of the disk drive which could be harvested by the given design.

4.4 Harvesting model

4.4.1 Governing equations

A piezoelectrical energy harvester, along with an energy storage device, is also modeled as a mass + damper + source + piezo structure [54,69,70]. It is made of a piezoelectric element that is attached to a mechanical structure that is linked to a storage circuit system. In this method, an efficient mass M under the forcing function $F(t)$ is confined to an efficient spring stiffness K, to a coefficient η damper, and to a piezoelectric element with an effective piezoelectric coefficient Θ and capacitance C_p. Consider, for instance, a triple layer bender mounted as a cantilever beam with polarization poled along the direction of thickness. Generation of the electric field is done by the thickness direction of the piezoelectric layers or the axial direction of strain. Modal analysis [71,72] can be used to derive the effective coefficients connected to the structural geometry and material constant:

$$M = \beta_M(m_p + m_b) + m_a, \tag{4.3}$$

$$K = \beta_K S\{(\frac{2}{3}\frac{t^3}{L^3} + \frac{ht^2}{L^3} + \frac{1}{2}\frac{th^2}{L^3})C_{p_{11}}^E + \frac{1}{12}\frac{h^3}{L^3}C_{b_{11}}^E, \tag{4.4}$$

$$\Theta = \beta_\Theta \frac{S(h+t)}{2L}e_{31}, \tag{4.5}$$

$$C_p = \frac{SL}{2t}\epsilon_{33}^S, \tag{4.6}$$

where the constants β_Θ, β_K, and β_M were derived from Rayleigh–Ritz approximation, piezoelectric constant is e_{31}, and ϵ_{33}^S is the clamped dielectric constant. Further, L is the axial length and S is the width of the cantilever beam. Thicknesses are denoted by t and h, elastic moduli by $C_{p_{11}}^E$ and $C_{b_{11}}^E$, while masses are given by m_p and m_b for piezoelectric and central passive layers, respectively. The attached mass is m_a. A less general, operated longitudinally or 3–3-mode, piezoelectric power generator is recently established in [73]. This mode has the advantage as the longitudinal piezoelectric effect is typically greater than the transverse effect ($d_{33} > d_{31}$).

A vibrating piezoelectric element produces an alternating current voltage, while an electrochemical battery needs a stabilized direct current voltage. To ensure electrical compatibility, an energy harvesting circuit is required. DC voltage is smoothed by using an AC–DC rectifier and a filtering capacitance C_e. For regulating the output voltage, a controller is placed between the battery and the rectifier output. Fig. 4.5 shows a simple har-

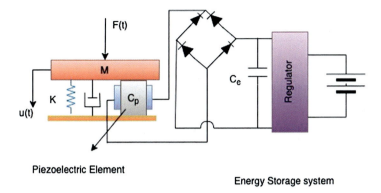

Figure 4.4 An equivalent model for a piezoelectric vibration energy harvesting system.

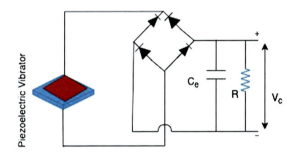

Figure 4.5 A typical AC–DC harvesting circuit.

vesting circuit. It should be noted that an equivalent resistor R replaced the battery and regulation circuit from Fig. 4.4, and V_c is the rectified voltage across it.

Displacement of the mass M is assumed to be u and the voltage across the piezoelectric element is denoted by V_p. The equations which govern the vibrator by conventional model analysis can be obtained as follows [74,75]:

$$M\ddot{u}(t) + \eta\dot{u}(t) + \Theta V_p(t) = F(t), \tag{4.7}$$

$$-\Theta\dot{u}(t) + C_p \dot{V}_p(t) = -I(t). \tag{4.8}$$

To the power generator, an AC–DC circuit is connected as shown in Fig. 4.5 where the flowing current through this circuit is represented by

72 Piezoelectric Aeroelastic Energy Harvesting

$I(t)$ and it is related to the rectified voltage V_c by

$$I(t) = \begin{cases} C_e \dot{V}_c(t) + \frac{V_c}{R} & \text{if } V_p = V_c, \\ -C_e \dot{V}_c(t) - \frac{V_c}{R} & \text{if } V_p = -V_c, \\ 0 & \text{if } |V_p| < V_c. \end{cases} \tag{4.9}$$

To the system a sinusoidal mechanical excitation is applied,

$$F(t) = F_0 \sin \omega t, \tag{4.10}$$

with constant magnitude F_0 and vibration's angular velocity ω (in rad s^{-1}). It must be noted that in the majority of the vibration-based power harvesters, exciting the base acceleration $\ddot{z}(t)$ is the reason for the system's source $F(t)$.

Explanation of Eq. (4.9) is as follows. If the voltage $|V_p|$ decreases below the rectified voltage V_c then the rectifying bridge will be an open circuit, which as a consequence vanishes the flowing current through the circuit. On the contrary, the bridge conducts if the voltage $|V_P|$ reaches V_c and the rectified voltage and the piezovoltage are kept equal, i.e., $|V_P| = V_c$. At the end when the piezovoltage $|V_P|$ value starts to decrease, an absolute value is reached which causes the block of conduction in the rectifier diodes.

A stable DC output voltage V_c is required by many applications, therefore to achieve this it can be assumed that the filter capacitance C_e is large enough such that the output voltage V_c becomes essentially constant [70].

Under steady-state operations, these equations can be solved by first determining the relation between rectified voltage's average value and the average value of displacement magnitude. If the rectifying bridge is blocked then the piezovoltage $V_p(t)$ varies proportionally to the displacement $u(t)$ with zero outgoing piezoelectric currents. Therefore the solutions for $V_p(t)$ and $u(t)$ can be assumed as

$$V_p(t) = g(\omega t - \theta), \tag{4.11}$$

$$u(t) = u_0 \sin(\omega t - \theta), \tag{4.12}$$

where constant displacement magnitude is represented by u_0 and a periodic function $g(t)$ has the period 2π and $|g(t)| \leq V_c$. Vibration period is assumed to be $T = \frac{2\pi}{\omega}$, where a and b two time instants ($b - a = \frac{T}{2}$) such that the displacement u takes values from the minimum of $-u_0$ to the maximum of u_0. During the semi-period from a to b, let us assume $\dot{V}_p \geq 0$. It follows that

$$\int_a^b \dot{V}_p(t)\,dt = V_c - (-V_c) = 2V_c. \tag{4.13}$$

It should be noted that for $a < t < t^*$, during which the piezovoltage $|V_p| < V_c$, we have $C_e \dot{V}_c(t) + \frac{V_c}{R} = 0$, and when $t^* \le t < b$ the rectifier conducts. Hence we have

$$\int_a^b T(t)\,dt = \frac{T}{2}\frac{V_c}{R}. \tag{4.14}$$

Since through the capacitance C_e the average current flowing is zero, for steady-state operation,

$$\int_a^b C_e \dot{V}_c(t)\,dt = 0. \tag{4.15}$$

Eq. (4.8) is integrated from time a to b, therefore we have

$$-2\Theta u_0 + 2C_p V_c = -\frac{T}{2}\frac{V_c}{R}, \tag{4.16}$$

$$V_c = \frac{\omega\Theta R}{\omega C_p R + \frac{\pi}{2}} u_0. \tag{4.17}$$

This equation is identical to that derived by the authors of [70,76]. In terms of the displacement magnitude, the average harvested power is

$$P = \frac{V_c^2}{R} = \frac{\omega^2 \Theta^2 R}{(\omega C_p R + \frac{\pi}{2})^2} u_0^2. \tag{4.18}$$

We must find u_0 to determine P and V_c. Estimation of this can be done by two approaches in the literature [70,76]. The analysis is simplified by introducing the following dimensionless parameters:

$$\omega_n = \sqrt{\frac{K}{M}}, \qquad k_e^2 = \frac{\Theta^2}{KC_p}, \qquad \zeta = \frac{\eta}{2\sqrt{KM}}, \tag{4.19}$$

$$\Omega = \frac{\omega}{\omega_n}, \qquad r = C_p \omega_n R. \tag{4.20}$$

Short circuit's natural frequency is ω_n, the alternative electromechanical coupling coefficient is represented by k_e^2, the damping ratio is ζ, r is the electric resistance, and the normalized frequency is denoted by Ω. The piezoelectric structure has both open circuit and short circuit stiffness, and there are two resonances of the system. They are defined by

$$\Omega_{sc} = 1, \qquad \Omega_{oc} = \sqrt{1 + k_e^2}, \qquad (4.21)$$

where Ω_{sc} is the frequency ratio of the short circuit and Ω_{oc} is the frequency ratio of the open circuit.

References

[1] Henry A. Sodano, Daniel J. Inman, Gyuhae Park, A review of power harvesting from vibration using piezoelectric materials, Shock and Vibration Digest 36 (3) (2004) 197–206.

[2] Mohsin Ali Marwat, Weigang Ma, Pengyuan Fan, Hassan Elahi, Chanatip Samart, Bo Nan, Hua Tan, David Salamon, Baohua Ye, Haibo Zhang, Ultrahigh energy density and thermal stability in sandwich-structured nanocomposites with dopamine@Ag@BaTiO3, Energy Storage Materials 31 (2020) 492–504.

[3] Ahsan Ali, Riffat Asim Pasha, Hassan Elahi, Muhammad Abdullah Sheeraz, Saima Bibi, Zain Ul Hassan, Marco Eugeni, Paolo Gaudenzi, Investigation of deformation in bimorph piezoelectric actuator: analytical, numerical and experimental approach, Integrated Ferroelectrics 201 (1) (2019) 94–109.

[4] Muhammad Usman Khan, Zubair Butt, Hassan Elahi, Waqas Asghar, Zulkarnain Abbas, Muhammad Shoaib, M. Anser Bashir, Deflection of coupled elasticity–electrostatic bimorph PVDF material: theoretical, FEM and experimental verification, Microsystem Technologies 25 (8) (2019) 3235–3242.

[5] Vittorio Memmolo, Hassan Elahi, Marco Eugeni, Ernesto Monaco, Fabrizio Ricci, Michele Pasquali, Paolo Gaudenzi, Experimental and numerical investigation of PZT response in composite structures with variable degradation levels, Journal of Materials Engineering and Performance 28 (6) (2019) 3239–3246.

[6] H. Elahi, A. Israr, R.F. Swati, H.M. Khan, A. Tamoor, Stability of piezoelectric material for suspension applications, in: 2017 Fifth International Conference on Aerospace Science & Engineering (ICASE), IEEE, 2017, pp. 1–5.

[7] Inderjit Chopra, Review of state of art of smart structures and integrated systems, AIAA Journal 40 (11) (2002) 2145–2187.

[8] Hassan Elahi, Marco Eugeni, Paolo Gaudenzi, Electromechanical degradation of piezoelectric patches, in: Analysis and Modelling of Advanced Structures and Smart Systems, Springer, 2018, pp. 35–44.

[9] Hassan Elahi, Khushboo Munir, Marco Eugeni, Muneeb Abrar, Asif Khan, Adeel Arshad, Paolo Gaudenzi, A review on applications of piezoelectric materials in aerospace industry, Integrated Ferroelectrics 211 (1) (2020) 25–44.

[10] Hassan Elahi, Marco Eugeni, Paolo Gaudenzi, Madiha Gul, Raees Fida Swati, Piezoelectric thermo electromechanical energy harvester for reconnaissance satellite structure, Microsystem Technologies 25 (2) (2019) 665–672.

[11] Hassan Elahi, Marco Eugeni, Paolo Gaudenzi, Faisal Qayyum, Raees Fida Swati, Hayat Muhammad Khan, Response of piezoelectric materials on thermomechanical shocking and electrical shocking for aerospace applications, Microsystem Technologies 24 (9) (2018) 3791–3798.

[12] Hassan Elahi, Zubair Butt, Marco Eugnei, Paolo Gaudenzi, Asif Israr, Effects of variable resistance on smart structures of cubic reconnaissance satellites in various thermal

and frequency shocking conditions, Journal of Mechanical Science and Technology 31 (9) (2017) 4151–4157.

[13] Hassan Elahi, M. Rizwan Mughal, Marco Eugeni, Faisal Qayyum, Asif Israr, Ahsan Ali, Khushboo Munir, Jaan Praks, Paolo Gaudenzi, Characterization and implementation of a piezoelectric energy harvester configuration: analytical, numerical and experimental approach, Integrated Ferroelectrics 212 (1) (2020) 39–60.

[14] Hassan Elahi, Khushboo Munir, Marco Eugeni, Sofiane Atek, Paolo Gaudenzi, Energy harvesting towards self-powered IoT devices, Energies 13 (21) (2020) 5528.

[15] Hassan Elahi, Marco Eugeni, Paolo Gaudenzi, A review on mechanisms for piezoelectric-based energy harvesters, Energies 11 (7) (2018) 1850.

[16] Zubair Butt, Riffat Asim Pasha, Faisal Qayyum, Zeeshan Anjum, Nasir Ahmad, Hassan Elahi, Generation of electrical energy using lead zirconate titanate (PZT-5a) piezoelectric material: analytical, numerical and experimental verifications, Journal of Mechanical Science and Technology 30 (8) (2016) 3553–3558.

[17] Shashank Priya, Advances in energy harvesting using low profile piezoelectric transducers, Journal of Electroceramics 19 (1) (2007) 167–184.

[18] Shad Roundy, Paul K. Wright, Jan Rabaey, A study of low level vibrations as a power source for wireless sensor nodes, Computer Communications 26 (11) (2003) 1131–1144.

[19] Shad Roundy, Paul K. Wright, A piezoelectric vibration based generator for wireless electronics, Smart Materials and Structures 13 (5) (2004) 1131.

[20] Shad Roundy, Eli S. Leland, Jessy Baker, Eric Carleton, Elizabeth Reilly, Elaine Lai, Brian Otis, Jan M. Rabaey, Paul K. Wright, V. Sundararajan, Improving power output for vibration-based energy scavengers, IEEE Pervasive Computing 4 (1) (2005) 28–36.

[21] Eli S. Leland, Paul K. Wright, Resonance tuning of piezoelectric vibration energy scavenging generators using compressive axial preload, Smart Materials and Structures 15 (5) (2006) 1413.

[22] Hassan Elahi, Khushboo Munir, Marco Eugeni, Paolo Gaudenzi, Reliability risk analysis for the aeroelastic piezoelectric energy harvesters, Integrated Ferroelectrics 212 (1) (2020) 156–169.

[23] Hassan Elahi, Marco Eugeni, Luca Lampani, Paolo Gaudenzi, Modeling and design of a piezoelectric nonlinear aeroelastic energy harvester, Integrated Ferroelectrics 211 (1) (2020) 132–151.

[24] Hassan Elahi, Marco Eugeni, Federico Fune, Luca Lampani, Franco Mastroddi, Giovanni Paolo Romano, Paolo Gaudenzi, Performance evaluation of a piezoelectric energy harvester based on flag-flutter, Micromachines 11 (10) (2020) 933.

[25] Hassan Elahi, The investigation on structural health monitoring of aerospace structures via piezoelectric aeroelastic energy harvesting, Microsystem Technologies (2020) 1–9.

[26] Marco Eugeni, Hassan Elahi, Federico Fune, Luca Lampani, Franco Mastroddi, Giovanni Paolo Romano, Paolo Gaudenzi, Numerical and experimental investigation of piezoelectric energy harvester based on flag-flutter, Aerospace Science and Technology 97 (2020) 105634.

[27] Hassan Elahi, Marco Eugeni, Paolo Gaudenzi, Design and performance evaluation of a piezoelectric aeroelastic energy harvester based on the limit cycle oscillation phenomenon, Acta Astronautica 157 (2019) 233–240.

[28] Marco Eugeni, Hassan Elahi, Federico Fune, Luca Lampani, Franco Mastroddi, Giovanni Paolo Romano, Paolo Gaudenzi, Experimental evaluation of piezoelectric energy harvester based on flag-flutter, in: Conference of the Italian Association of Theoretical and Applied Mechanics, Springer, 2019, pp. 807–816.

[29] Henry A. Sodano, G.H. Park, Donald J. Leo, Daniel J. Inman, Electric power harvesting using piezoelectric materials, Center for Intelligent Material Systems and Structures, Virginia Polytechnic Institute and State University, 2003.

76 Piezoelectric Aeroelastic Energy Harvesting

[30] Henry A. Sodano, Daniel J. Inman, Gyuhae Park, Comparison of piezoelectric energy harvesting devices for recharging batteries, Journal of Intelligent Material Systems and Structures 16 (10) (2005) 799–807.

[31] Dongna Shen, Jung-Hyun Park, Joo Hyon Noh, Song-Yul Choe, Seung-Hyun Kim, Howard C. Wikle III, Dong-Joo Kim, Micromachined PZT cantilever based on SOI structure for low frequency vibration energy harvesting, Sensors and Actuators A: Physical 154 (1) (2009) 103–108.

[32] Jing-Quan Liu, Hua-Bin Fang, Zheng-Yi Xu, Xin-Hui Mao, Xiu-Cheng Shen, Di Chen, Hang Liao, Bing-Chu Cai, A MEMS-based piezoelectric power generator array for vibration energy harvesting, Microelectronics Journal 39 (5) (2008) 802–806.

[33] W.J. Choi, Yongbae Jeon, J.-H. Jeong, Rajendra Sood, Sang-Gook Kim, Energy harvesting MEMS device based on thin film piezoelectric cantilevers, Journal of Electroceramics 17 (2–4) (2006) 543–548.

[34] Hyeoung Woo Kim, Shashank Priya, Kenji Uchino, Robert E. Newnham, Piezoelectric energy harvesting under high pre-stressed cyclic vibrations, Journal of Electroceramics 15 (1) (2005) 27–34.

[35] Xiaotian Li, Mingsen Guo, Shuxiang Dong, A flex-compressive-mode piezoelectric transducer for mechanical vibration/strain energy harvesting, IEEE Transactions on Ultrasonics, Ferroelectrics, and Frequency Control 58 (4) (2011) 698–703.

[36] Anita M. Flynn, Seth R. Sanders, Fundamental limits on energy transfer and circuit considerations for piezoelectric transformers, IEEE Transactions on Power Electronics 17 (1) (2002) 8–14.

[37] Niell G. Elvin, Alex A. Elvin, Myron Spector, A self-powered mechanical strain energy sensor, Smart Materials and structures 10 (2) (2001) 293.

[38] Alper Erturk, Daniel J. Inman, Issues in mathematical modeling of piezoelectric energy harvesters, Smart Materials and Structures 17 (6) (2008) 065016.

[39] Alper Erturk, Daniel J. Inman, On mechanical modeling of cantilevered piezoelectric vibration energy harvesters, Journal of Intelligent Material Systems and Structures 19 (11) (2008) 1311–1325.

[40] Alper Erturk, Daniel J. Inman, An experimentally validated bimorph cantilever model for piezoelectric energy harvesting from base excitations, Smart Materials and Structures 18 (2) (2009) 025009.

[41] Carlos De Marqui Junior, Alper Erturk, Daniel J. Inman, An electromechanical finite element model for piezoelectric energy harvester plates, Journal of Sound and Vibration 327 (1–2) (2009) 9–25.

[42] Jamil M. Renno, Mohammed F. Daqaq, Daniel J. Inman, On the optimal energy harvesting from a vibration source, Journal of Sound and Vibration 320 (1–2) (2009) 386–405.

[43] G. Poulin, E. Sarraute, F. Costa, Generation of electrical energy for portable devices: comparative study of an electromagnetic and a piezoelectric system, Sensors and Actuators A: physical 116 (3) (2004) 461–471.

[44] Jyoti Ajitsaria, Song-Yul Choe, D. Shen, D.J. Kim, Modeling and analysis of a bimorph piezoelectric cantilever beam for voltage generation, Smart Materials and Structures 16 (2) (2007) 447.

[45] Yuantai Hu, Ting Hu, Qing Jiang, Coupled analysis for the harvesting structure and the modulating circuit in a piezoelectric bimorph energy harvester, Acta Mechanica Solida Sinica 20 (4) (2007) 296–308.

[46] Yi-Chung Shu, I.C. Lien, Efficiency of energy conversion for a piezoelectric power harvesting system, Journal of Micromechanics and Microengineering 16 (11) (2006) 2429.

[47] Marcin Marzencki, Maxime Defosseux, Skandar Basrour, MEMS vibration energy harvesting devices with passive resonance frequency adaptation capability, Journal of Microelectromechanical Systems 18 (6) (2009) 1444–1453.

[48] J.M. Dietl, A.M. Wickenheiser, E. Garcia, A Timoshenko beam model for cantilevered piezoelectric energy harvesters, Smart Materials and Structures 19 (5) (2010) 055018.

[49] John M. Dietl, Ephrahim Garcia, Beam shape optimization for power harvesting, Journal of Intelligent Material Systems and Structures 21 (6) (2010) 633–646.

[50] Luca Gammaitoni, Igor Neri, Helios Vocca, The benefits of noise and nonlinearity: extracting energy from random vibrations, Chemical Physics 375 (2–3) (2010) 435–438.

[51] Xiaotong Gao, Wei-Heng Shih, Wan Y. Shih, Vibration energy harvesting using piezoelectric unimorph cantilevers with unequal piezoelectric and nonpiezoelectric lengths, Applied Physics Letters 97 (23) (2010) 233503.

[52] Ryan R. Knight, Changki Mo, William W. Clark, MEMS interdigitated electrode pattern optimization for a unimorph piezoelectric beam, Journal of electroceramics 26 (1–4) (2011) 14–22.

[53] R. Ly, M. Rguiti, S. D'astorg, A. Hajjaji, C. Courtois, A. Leriche, Modeling and characterization of piezoelectric cantilever bending sensor for energy harvesting, Sensors and Actuators A: Physical 168 (1) (2011) 95–100.

[54] Cecilia D. Richards, Michael J. Anderson, David F. Bahr, Robert F. Richards, Efficiency of energy conversion for devices containing a piezoelectric component, Journal of Micromechanics and Microengineering 14 (5) (2004) 717.

[55] S-J. Jeong, D-S. Lee, M-S. Kim, D-H. Im, I-S. Kim, K-H. Cho, Properties of piezoelectric ceramic with textured structure for energy harvesting, Ceramics International 38 (2012) S369–S372.

[56] R. Elfrink, T.M. Kamel, M. Goedbloed, S. Matova, D. Hohlfeld, Y. Van Andel, R. Van Schaijk, Vibration energy harvesting with aluminum nitride-based piezoelectric devices, Journal of Micromechanics and Microengineering 19 (9) (2009) 094005.

[57] Cam Minh Tri Tien, Nam Seo Goo, Use of a piezocomposite generating element in energy harvesting, Journal of Intelligent Material Systems and Structures 21 (14) (2010) 1427–1436.

[58] Mateu Loreto, Francesc Moll, Optimum piezoelectric bending beam structures for energy harvesting using shoe inserts, Journal of Intelligent Material Systems and Structures 16 (10) (2005) 835–845.

[59] Mateu Loreto, Francesc Moll, Appropriate charge control of the storage capacitor in a piezoelectric energy harvesting device for discontinuous load operation, Sensors and Actuators A: Physical 132 (1) (2006) 302–310.

[60] Jonathan Granstrom, Joel Feenstra, Henry A. Sodano, Kevin Farinholt, Energy harvesting from a backpack instrumented with piezoelectric shoulder straps, Smart Materials and Structures 16 (5) (2007) 1810.

[61] Kevin M. Farinholt, Nicholas A. Pedrazas, David M. Schluneker, David W. Burt, Charles R. Farrar, An energy harvesting comparison of piezoelectric and ionically conductive polymers, Journal of Intelligent Material Systems and Structures 20 (5) (2009) 633–642.

[62] I. Patel, E. Siores, T. Shah, Utilisation of smart polymers and ceramic based piezoelectric materials for scavenging wasted energy, Sensors and Actuators A: Physical 159 (2) (2010) 213–218.

[63] J.W. Sohn, Seung B. Choi, D.Y. Lee, An investigation on piezoelectric energy harvesting for MEMS power sources, Proceedings of the Institution of Mechanical Engineers, Part C: Journal of Mechanical Engineering Science 219 (4) (2005) 429–436.

[64] Hongping Hu, Chun Zhao, Shengyuan Feng, Yuantai Hu, Chuanyao Chen, Adjusting the resonant frequency of a PVDF bimorph power harvester through a corrugation-shaped harvesting structure, IEEE transactions on ultrasonics, ferroelectrics, and frequency control 55 (3) (2008) 668–674.

[65] Yiming Liu, Geng Tian, Yong Wang, Junhong Lin, Qiming Zhang, Heath F. Hofmann, Active piezoelectric energy harvesting: general principle and experimental demonstration, Journal of Intelligent Material Systems and Structures 20 (5) (2009) 575–585.

[66] Chieh Chang, Van H. Tran, Junbo Wang, Yiin-Kuen Fuh, Liwei Lin, Direct-write piezoelectric polymeric nanogenerator with high energy conversion efficiency, Nano Letters 10 (2) (2010) 726–731.

[67] Benjamin J. Hansen, Ying Liu, Rusen Yang, Zhong Lin Wang, Hybrid nanogenerator for concurrently harvesting biomechanical and biochemical energy, ACS Nano 4 (7) (2010) 3647–3652.

[68] Jen-Yuan Chang, Modeling and analysis of piezo–elastica energy harvester in computer hard disk drives, IEEE Transactions on Magnetics 47 (7) (2011) 1862–1867.

[69] Elie Lefeuvre, Adrien Badel, Claude Richard, Daniel Guyomar, Piezoelectric energy harvesting device optimization by synchronous electric charge extraction, Journal of Intelligent Material Systems and Structures 16 (10) (2005) 865–876.

[70] Geffrey K. Ottman, Heath F. Hofmann, Archin C. Bhatt, George A. Lesieutre, Adaptive piezoelectric energy harvesting circuit for wireless remote power supply, IEEE Transactions on Power Electronics 17 (5) (2002) 669–676.

[71] Nesbitt W. Hagood, Walter H. Chung, Andreas Von Flotow, Modelling of piezoelectric actuator dynamics for active structural control, Journal of Intelligent Material Systems and Structures 1 (3) (1990) 327–354.

[72] Qing-Ming Wang, L. Eric Cross, Constitutive equations of symmetrical triple layer piezoelectric benders, IEEE transactions on ultrasonics, ferroelectrics, and frequency control 46 (6) (1999) 1343–1351.

[73] Y.B. Jeon, R. Sood, J-H. Jeong, S-G. Kim, MEMS power generator with transverse mode thin film PZT, Sensors and Actuators A: Physical 122 (1) (2005) 16–22.

[74] Noël E. Dutoit, Brian L. Wardle, Sang-Gook Kim, Design considerations for MEMS-scale piezoelectric mechanical vibration energy harvesters, Integrated Ferroelectrics 71 (1) (2005) 121–160.

[75] Henry A. Sodano, Gyuhae Park, D.J. Inman, Estimation of electric charge output for piezoelectric energy harvesting, Strain 40 (2) (2004) 49–58.

[76] Daniel Guyomar, Adrien Badel, Elie Lefeuvre, Claude Richard, Toward energy harvesting using active materials and conversion improvement by nonlinear processing, IEEE transactions on ultrasonics, ferroelectrics, and frequency control 52 (4) (2005) 584–595.

CHAPTER 5

Energy harvesting and circuits

Contents

5.1.	Introduction	79
5.2.	Piezoelectric energy harvesting circuits	81
5.3.	Energy conditioning circuits	82
	5.3.1 Synchronous circuits	82
	5.3.2 Circuitry for enhanced energy harvesting	84
5.4.	Equivalent circuit method	85
	5.4.1 Design of piezoelectric systems	86
	5.4.2 Power calculation for piezoelectric energy harvesting	86
5.5.	Impedance method circuit	87
	5.5.1 MPPT for piezoelectricity	88
	5.5.2 SSHI and SECE	91
	5.5.3 Rectifiers	91
	5.5.4 Voltage doublers	93
References		93

5.1 Introduction

The process of generating and collecting energy from external energy sources (e.g., solar, thermal energy, wind power, salinity gradient, and kinetic energy, also known as ambient energy) captures and stores energy for small, automatic, wireless equipment, such as portable electronics, and the network of wireless sensors is a process through which energy is collected. Among them piezoeletric materials are playing a vital role [1–4]. Currently, energy harvesting is mainly contributing to low-power electronics and it is a source of clean energy, which is very important from the environmental point of view [5,6]. Moreover, for the high-power generation methods, somehow the environment is also compromised and is expensive as well. For example, the fuel sources for high-power generation are oil and coal, whereas the sources for energy harvesting are its surroundings and atmosphere [7,8].

A major interest has been seen in energy harvesting and storage systems that convert atmospheric energy into electrical energy, especially in the military and industrial applications [9,10]. For example, one is interested in an instrument that converts atmospheric energy into electrical energy, such

Piezoelectric Aeroelastic Energy Harvesting
https://doi.org/10.1016/B978-0-12-823968-1.00016-7

Copyright © 2022 Elsevier Inc.
All rights reserved.

79

as ocean waves into energy used by autonomous oceanographic sensors. Future implementation will include the installation of high–performance equipment (or clusters) at remote sites as powerful power plants for large systems. Wearable electronics are also used for tracking or loading mobile phones, personal computers, radio networking equipment, etc., for energy storage systems. These instruments must be durable enough for long-term exposure to extreme conditions and be dynamically adaptive to take advantage of the full continuum of wave movements [11,12].

The history of energy generation and storage dates back to the windmill and the water wheel. For many decades, people have been searching for ways to store energy from heat and vibration energy. One of the driving factors behind the quest for new energy storage devices is the need to power sensor networks and mobile devices without batteries. The desire to fix climate change and the problem of global warming often motivates energy harvesting.

Small autonomous sensors like those built using microelectromechanical systems (MEMS) technology can also be used for power harvesting electricity. These devices are often very compact and require low power, but the dependence on battery power restricts their applications. The availability of smart sensors can result in scaling up of the ambient vibration capacity, wind, temperature or illumination [13,14].

The standard electricity densities available in power collectors are highly dependent on the particular application (which influences the size of the generator) and the configuration of the collector itself. Typical values for sensorized systems are usually of a few $\frac{\mu W}{cm^3}$ for applications of motion control [15,16] and hundreds or thousands of $\frac{\mu W}{cm^3}$ for machinery generators. Many energy scavengers produce very little power for wearable electronics.

After energy is generated, it can be used directly or may usually be contained in a capacitor, or a battery, or a supercapacitor. When the program requires massive energy surges, capacitors are used. When the system needs to provide a constant energy supply, batteries are used because of the steady energy flow. Supercapacitors have nearly infinite charge–discharge periods compared with batteries and can thus still run in order for the IoT and wireless sensor systems to operate maintenance-free.

Low energy harvesting interests currently apply to independent sensor networks. In such applications, the energy harvesting scheme carries the energy captured in the capacitor and then boosted/regulated to the second storage capacitor or battery for use in the microprocessor or data transmis-

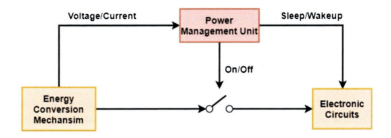

Figure 5.1 Energy harvesting circuit that utilizes energy on the spot.

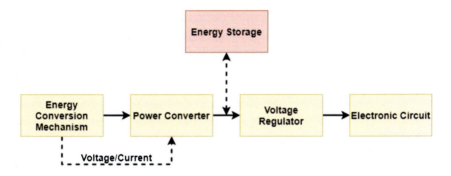

Figure 5.2 Energy harvesting circuit that stores energy.

sion. Power is usually used in a sensor application and the data stored or possibly transmitted by a wireless method. Below we present the common mechanisms for low-power energy harvesting mechanisms.

5.2 Piezoelectric energy harvesting circuits

Circuits play a vital role in the energy harvesting phenomenon since they can control the flow, avoid circuit breakage, regulate voltage, amplify the current, convert AC into DC, or vice versa, etc. In this chapter, we will only focus on energy harvesting circuits related to piezoelectric energy harvesting. In the literature, basically, two different types of circuit are used in energy harvesting mechanisms: one circuit can generate electric voltage and utilize it at the same time to drive electronic equipment as shown in Fig. 5.1; the second circuit stores the energy generated by the harvester and can utilize it when it is needed as shown in Fig. 5.2.

5.3 Energy conditioning circuits

Piezoelectric-powered systems harvest alternative voltages or AC power by using ambient noises to excite the retriever harmonically. Before using any appliances or storage devices that require DC power, this AC power must be conditioned/converted/rectified. There are two major phases of a standard power harvesting circuit: conditioning (AC/DC transformation) and controlling (DC/DC transformation). The simplest energy harvesting circuit is the total wave rectifier coupled with a smoothing condenser (to convert the signal to DC). But it is inefficient for converting extracted energy into stored energy and does not control voltage, which is the downside of this circuit—it lacks optimization. Advanced circuits with optimum tuning circuitry, sleep modes, degrading functionality, over-voltage battery safety and more are used widely in the literature [17].

In order to present a sufficient voltage at load, most energy harvesting circuits are using some form of voltage control after rectification. There are many DC–DC converter topologies to be used for this purpose, based on the relative voltage of the harvester output and the desired value of load input. If the harvester voltage output is higher than the load specifications for voltage, a linear voltage control system is the simplest circuit. While different researchers use linear regulators, they are generally ineffective [18]. An alternative way to achieve a safer approach is to use a DC–DC converter, called a step-down converter or buck converter [19].

In the case of a less than the load voltage demand, a step-up converter or boost converter may be used if the voltage output of the harvester is not adequate. Even more, there can be times where the harvester voltage is variable in time, at a given time, and maybe higher or lower than the load voltage necessary. If so, a buck–boost converter that incorporates both the buck and boost converter's features can be used [20].

5.3.1 Synchronous circuits

Although switching DC–DC converters can be used in an energy collection system for the necessary voltage control, their performance remains suboptimal. In order to develop basic switching converters, Lefeuvre et al. presented an important idea in the form of the SECE which is a synchronous electric charge extraction concept [21]. The SECE concept involves synchronization of the energy extracted from the piezoelectric and delivered to the load with the maxima and minima of the harvester's dis-

Table 5.1 Topologies and synchronized switch techniques for energy harvesting.

Techniques	Abbreviation	Year	Ref.
Synchronous-electric-charge-extraction	SECE	2005	[21]
Synchronized-switch-harvesting on inductor	SSHI	2005	[22,23]
Series-synchronized-switch-harvesting on inductor	SSHI	2006	[24]
Double-synchronized-switch-harvesting	DSSH	2008	[25]
Enhanced-synchronized-switch-harvesting	ESSH	2009	[27]
Synchronized-switch-harvesting on inductor using magnetic rectifier	SSHI-MR	2009	[28]
Hybrid-synchronized-switch-harvesting on inductor	HSSHI	2011	[29]
Self-powered-synchronized-switch-harvesting on inductor	SP-SSHI	2012	[30]
Frequency-tuning-synchronized-charge extraction	FTSECE	2016	[31]
Unipolar-synchronized-charge extraction	USECE	2018	[32]
Synchronized-triple bias-flip	P-S3BF	2019	[33]

placement (which corresponds to generated voltage). In their work, the circuit topology had revealed that the transfer of energy relative to the unsynchronized energy harvesting was increased by four times. The same research group continued to incorporate the coordinated selection cycle principle after this initial work. The methodology synchronized switch harvesting on inductor (SSHI) is named after Guyomar et al. [22] and Badel et al. [23] who used a circuit inductor between the harvester and rectifier.

The SSHI series topology technique was proposed by Lefeuvre et al. [24] in which the switch and the inducer were placed in series with the generator before the rectifier. The findings indicated that energy transmission was improved by 15 times compared with direct charging; both series and parallel SSHI topologies were tested in their research. Additional adaptations to the synchronized switching principle are rendered by the DSSH proposed by Lallart et al. [25]. The major portion of circuitry study for energy harvesting takes into consideration synchronization techniques. Chao presented a description of all these works related to the synchronized circuitry in detail [26]. Topologies techniques for energy harvesting are represented in Table 5.1.

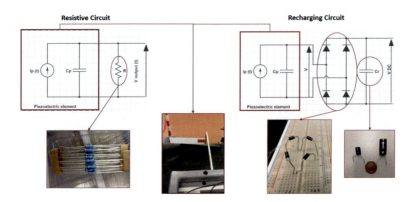

Figure 5.3 Aeroelastic piezoelectric energy harvesting circuits [37].

5.3.2 Circuitry for enhanced energy harvesting

Impedance matching is another method used for energy collection circuits to maximize the performance of the harvesting process. If the impedances of a source and load are the same then maximum harvesting can be achieved [34]. Elahi et al. [35–40] used resistive and capacitive circuits (resistive circuit for direct energy harvesting and capacitive circuit to store energy) for the aeroelastic piezoelectric energy harvesting as shown in Fig. 5.3. Kong et al. have developed a matching impedance circuit that incorporates regular diode rectifiers with discontinuous conduction buck–boost converters [41]. To adjust the effective impedance of the piezoelectric harvester, the duty cycle of the switching buck–boost converter may be adjusted. The method of impedance matching was also used by Kim et al. in the buck converter topology [42]. Guyomar et al. explored the idea of achieving impedance balancing using the previously mentioned synchronous circuit architecture [43]. The DSSH system incorporates a front-end of the sequence SSHI with a back-end buck–boost. By changing the capacitor values in the circuit, the effective impedance of the harvest circuit can be calculated. The results of the study showed that the DSSH technology could increase performance five times compared with direct charge [29].

Improving the processes and circuits of piezoelectric energy selection, including SECE, SSHI, and impedance compatibility, is an active area of research. A new FTSECE technique has been defined briefly to enhance the performance of SECE circuits for resonator systems with high electromechanical connections [31]. This approach eliminates the two problems associated with standard SECE techniques, namely the narrow frequency bandwidth in heavily coupled generators and the impairment when regulating the voltage decline when the generator is removed.

This approach has been used by Brenes et al. in the experimental set-up [44]. In comparison to the conventional SECE method, it has shown a substantial improvement of the harvester's frequency bandwidth. In addition, a shunt diode for low amplitude vibration and low piezoelectric voltage (less than 2 VRMS) applications was used as a bridge rectifier traditionally used in SECE circuits [32]. The process was referred to as USECE. Experimental studies have shown that the combination of SECE's high input capacity with unipolar operating power efficiency provides 75% of power, while the standard SECE circuit provides a power efficiency of less than 35%. Moreover, relative to a SECE load-conditioning circuit, the USECE circuit reported a 200% improvement in the power output.

In addition to the impedance matching circuit designed for generators with strong electromechanical coupling, Brenes et al. implemented the shunt-diode approach in piezoelectric energy harvesting [45]. In order to provide the unidirectional voltage signal to a DC–DC converter for power optimization, the shunt diode can be attached directly to the piezoelectric transducer. Although SSHI circuits boost the efficiency of weakly connected piezoelectric harvesters, SSHI is less effective because of comparatively consistent bandwidth as the electromechanical connectivity increases. It is demonstrated that the shunt diode architecture could provide a greater bandwidth in this case compared to the SSHI. The architecture proposed also demonstrated better performance in comparison to AC–DC transducers for low piezoelectric power outputs, which reduces the efficiency for low voltage levels.

Zhao et al. have developed a new P-S3BF technique to minimize the energy dissipation on SECE and SSHI circuits in passive voltage bias–flip operations [33]. Experimental tests on piezoelectric cantilevers have found that the SSHI is 287.6% higher than normal bridge rectifier circuits and produces 24.5% more energy than SSHI.

5.4 Equivalent circuit method

The dynamics of piezoelectric structures can be modeled using electromechanical equivalent circuits. These are applied to the modeling of piezoelectric generators for the scavenging of electric potential [46]. The emphasis in the early stages is on the mathematical modeling and low-complexity device depictions. Subsequently, comprehensive models would be needed,

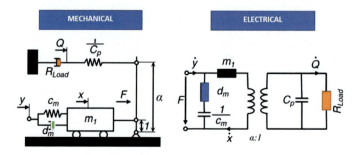

Figure 5.4 Equivalent circuit method for piezoelectric energy harvester.

which take into account all applicable design parameters. In general experiments, early in the design process, piezoelectric circuit equivalents are commonly used. Many researchers have used equivalent circuit methods for the prediction of degradation in piezoelectric materials [47,48].

5.4.1 Design of piezoelectric systems

For the study and design of piezoelectric systems, equivalent circuit models may be used. Various techniques are used to achieve the parameters in these models. These approaches include the experimental recognition parameter, focused on the transfer functions of the devices. They can be applied anywhere there is a physical prototype. It applies even if numerical calculations are used to extract data from the transfer function. Prototypes were usually not available in the early design stages when the overall system is synthesized. In this case, some empirical solutions of the Euler–Bernoulli beam equation can be used to determine the model parameters.

5.4.2 Power calculation for piezoelectric energy harvesting

In Fig. 5.4, equivalent circuit method for piezoelectric energy harvester is represented in mechanical and electrical form, respectively. Dielectric losses are assumed to be neglected and the load (electrical) R_{Load} will be modeled on a resistive load. From the equations of motion, we get

$$c_m(x(t) - y(t)) + d_m(\dot{x}(t) - \dot{y}(t)) + m_1 \ddot{x}(t) = -\alpha F(t), \quad (5.1)$$

$$1/C_p(\alpha x(t) - Q(t)) = F(t), \quad (5.2)$$

$$R_{\text{Load}} \dot{Q}(t) = F(t). \quad (5.3)$$

Optimal power output is one of the most critical aspects of an energy harvesting circuit. The voltage amplitude at output is represented in Eq. (5.5) and the output power is given in Eq. (5.6). From Eq. (5.6), it is observed that when the load resistance matches the impedance of the harvester, optimum power transmission is obtained. The equations are as follows:

$$P(j\omega) = U_L(j\omega)i^*(j\omega),$$ (5.4)

$$U_L(j\omega) = \frac{(c_m + j\omega d_m)\alpha R_{\text{Load}}\dot{y}}{j\omega R_{\text{Load}}\alpha^2 + (c_m + j\omega(d_m + j\omega m_1))(1 + j\omega C_p R_{\text{Load}})},$$ (5.5)

$$P(j\omega) = R_{\text{Load}}(\frac{U_L(j\omega)}{R_{\text{Load}}})^*.$$ (5.6)

5.5 Impedance method circuit

When the load impedance is the complex conjugate of the source impedance, the maximal power is transmitted. Since a piezoelectric generator has a wide capacity, complex conjugation needs a broad inductance to cancel the load. In addition, the value of inductance depends on the frequency of vibration. Consequently, the use of a passive inductor is not feasible. Saggini et al. analyzed the dynamic mixture method that imitates an inductive system [49]. The input voltage was greater than the output voltage, and they used a bidirectional DC/DC converter. The current thus flows under the buck mode from the piezoelectric generator to the battery and in the opposite direction in boosting mode. Their solution leads to the dissipation of significant power by a dynamic loop. Resistive matching is an alternative approach to complex conjugate matching. The source impedance Z_s can be defined as

$$Z_s = R_s + jX_s,$$ (5.7)

where the source resistance is R_s and the excitation frequency of the harvester is X_s. Then, the optimal resistance R_L can be represented as [41]

$$R_L = \sqrt{R_s^2 + X_s^2}.$$ (5.8)

From Eq. (5.8), it can be observed that the optimal resistance is also dependent on the harvester's excitation frequency. Therefore, to compensate for it, the load resistance can be varied as well, and this phenomenon is known as maximum power point tracking (MPPT). The cost of resistant matching is that the highest energy is only transmitted to the load at the resonant frequency compared to complex conjugate. At other frequencies, which are

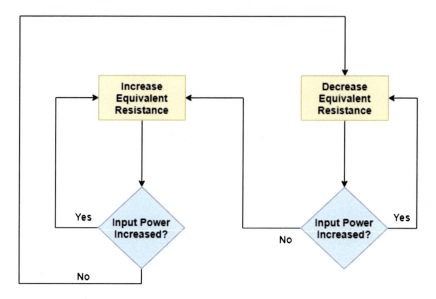

Figure 5.5 Hill climbing algorithm for energy harvesting.

away from the resonant frequency, the power level decreases dramatically [50].

5.5.1 MPPT for piezoelectricity

When the operational state involves MPPT modifications, the source impedance of a transducer has the piezoelectric generator and a photovoltaic cell. The use of an MPPT scheme would only be beneficial if the extra power received due to the MPPT is more than that discharged by the MPPT circuit itself. There are limits on the additional power that can be harvested by MPPT, the transducer and its working state. The precision and the speed of tracking, sometimes exchanged, is another architecture problem with MPPT. A more precise and thus more complex MPPT system often dissipates greater power for the MPPT. There is therefore no need to improve the net energy obtained by the implementation of an accurate system.

Hill climbing and fractional open-circuit voltage (FOCV) are two widely used MPPT algorithms for energy selection. Fig. 5.5 demonstrates the algorithm for maximum energy harvesting, also called disturbing and observing, which shows the complex resistive match. It raises the resistance to load by a certain step and calculates the power supplied to the load, always the average current. As power increases, the course of the quest is right, and in the next step, it increases resistance again. When the power falls, the trajectory is reversed. If there is no local limit, the algorithm implies that convergence is maximal, as is the case for piezoelectric generators. With an MCU or a mixed-signal circuit, the climbing algorithm can be applied [50,51]. In consideration of the nearly equal voltage of a storage unit per cycle, the power can be determined in the storage system by calculating the average electricity.

A flyback converter for dynamic resistive matching has been integrated with the proposed power management system for piezoelectric energy generation. The proposed system is adaptable continuously and can start cold without a backup battery. The system can start cold. For the powerpoint, a DCM flyback converter with a continuous on-time modulation is used. To optimize energy collection, the programming of a low-power microcontrol unit (MCU) provides optimum power point monitoring (MPPT). In general, the MCU is still in a WSN. The use of MCU thus needs no external hardware. Other types of energy storage system, including small wind turbines and solar panels, are also readily feasible with the low-power architecture features [50].

The FOCV algorithm is based on the assumption that the voltage over the load is 1/2 of the source's open circuit stress [52]. Compared to the hill climbing algorithm, the FOCV algorithm is easier to use. But, for calculating the open-circuit voltage, the FOCV algorithm allows the load to be open once in a while and the energy harvesting to be disturbed throughout the time. Fig. 5.6 demonstrates the algorithm for maximum energy harvesting called FOCV. The benefit of this approach is that it needs only one sensor to control the stress, and the device load is much smaller than in other normal and improvised algorithms while computing [53]. The consequences of the MPPT performance reduction in the traditional FOCV mechanism can be overcome by the difference of panel open-circuit voltage due to variations in temperature. Moreover, experimental findings revealed a higher tracking performance compared with standard algorithms for the proposed IFOCV algorithm. The test results were checked even by the analysis of MPPT output obtained by the traditional FOCV algorithm with the IFOCV proposed. Compared to other MPPT algorithms, the IFOCV algorithm is also much simpler, more effective and computer-intensive [53].

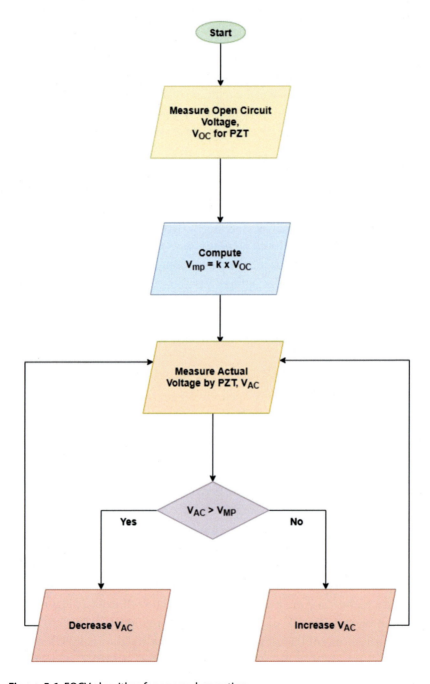

Figure 5.6 FOCV algorithm for energy harvesting.

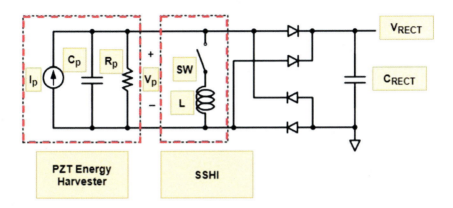

Figure 5.7 SSHI circuit for a piezoelectric energy harvester.

5.5.2 SSHI and SECE

Synchronized switch harvesting on inductor (SSHI) targets to remove the influence of the piezoelectric generator capacitive parameters [22,54,55]. The SSHI circuit consisting of inductor L and switch S is shown in Fig. 5.7. A simpler equivalent Norton circuit for the PZT model is represented in Fig. 5.7. The PZT condenser C_p is filled negatively from negative to positive at the crossing point of the PZT current I_p.

The SSHI device eliminates the excess energy used to positively charge the condenser using an LC resonant system. After the zero crossings of PZT current i_p, the SW switch is locked for a moment. The condenser C_p and the inductor L form a resonant circuit with a resonant frequency far higher than the PZT vibrating frequency. If the resonant circuit oscillates, the condenser voltage V_P increases rapidly and the switch is opened at the opposite peak value of V_P, reducing the energy consumption of the PZT to discharge the capacitor. As the PZT current I_p shifts from positive to negative, the same procedure is carried out to conserve energy to negatively fill the capacitor. The energy extracted from the SSHI circuit increases relative to a solution circuit just at the cost of greater complexity.

5.5.3 Rectifiers

To transform AC voltage output into DC, piezoelectric generators need to be rectified. Although a complete flange correction system can be integrated with passive diodes, the reduction in front of the diode may cause considerable power loss. Active or synchronous rectifiers minimize passive

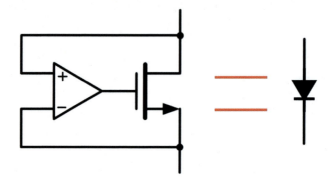

Figure 5.8 Rectification circuit for a piezoelectric energy harvester.

diode errors. The rectification circuit for the piezoelectric energy harvester is shown in Fig. 5.8. In reasonable time intervals, an active rectifier activates or disables an MOSFET to reliably rectify the voltage [56–58].

Fig. 5.8 demonstrates an active rectification based on a comparator that detects the input voltage polarity while triggering or disabling the MOSFET. If the flow is small through MOSFET, the disparity in the drain and spring stress are small near zero point crossings to contribute to a failure. Different circuit designs were proposed for corrections, and a key design concern is the lower dissipation of power of the gate control system, i.e., the comparator [59–61]. The piezoelectric generator also has a problem that the rectified voltage is smaller, resulting in low boost converter's efficiency. Therefore, there is a need for a circuit that can enhance the amplitude of the rectified voltage as well.

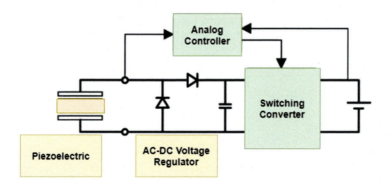

Figure 5.9 Schematics of the circuit for voltage doublers.

5.5.4 Voltage doublers

As discussed about rectifiers, the rectified voltage is smaller, this issue can be overcome by using voltage doublers [19]. The proposed circuit for voltage doublers is represented in Fig. 5.9. The output voltage increases twice with rectification by a voltage doubler. The procedure of a voltage doubler and active diodes can be merged for better results. Moreover, in contrast to the conventional rectification circuit, the reconfigurable rectifying device can transfer electrical energy to the loading resistor four times higher [59].

References

[1] G. Park, H. Sohn, C.R. Farrar, D.J. Inman, Overview of piezoelectric impedance-based health monitoring and path forward, Shock and Vibration Digest 35 (6) (2003) 451–464.

[2] J.J. Dosch, D.J. Inman, E. Gracia, A self-sensing piezoelectric actuator for collocated control, Journal of Intelligent Material Systems and Structures 3 (1) (1992) 166–185.

[3] G.K. Ottman, H.F. Hofmann, A.C. Bhatt, G.A. Lesieutre, Adaptive piezoelectric energy harvesting circuit for wireless remote power supply, IEEE Transactions on Power Electronics 17 (5) (2002) 669–676.

[4] H.A. Sodano, D.J. Inman, G.H. Park, A review of power harvesting from vibration using piezoelectric materials, IEEE Transactions on Power Electronics 36 (3) (2004) 197–206.

[5] Hassan Elahi, M. Rizwan Mughal, Marco Eugeni, Faisal Qayyum, Asif Israr, Ahsan Ali, Khushboo Munir, Jaan Praks, Paolo Gaudenzi, Characterization and implementation of a piezoelectric energy harvester configuration: analytical, numerical and experimental approach, Integrated Ferroelectrics 212 (1) (2020) 39–60.

[6] Hassan Elahi, Khushboo Munir, Marco Eugeni, Sofiane Atek, Paolo Gaudenzi, Energy harvesting towards self-powered IoT devices, Energies 13 (21) (2020) 5528.

[7] Hassan Elahi, Marco Eugeni, Paolo Gaudenzi, A review on mechanisms for piezoelectric-based energy harvesters, Energies 11 (7) (2018) 1850.

[8] Zubair Butt, Riffat Asim Pasha, Faisal Qayyum, Zeeshan Anjum, Nasir Ahmad, Hassan Elahi, Generation of electrical energy using lead zirconate titanate (PZT-5a) piezoelectric material: analytical, numerical and experimental verifications, Journal of Mechanical Science and Technology 30 (8) (2016) 3553–3558.

[9] Hassan Elahi, Khushboo Munir, Marco Eugeni, Muneeb Abrar, Asif Khan, Adeel Arshad, Paolo Gaudenzi, A review on applications of piezoelectric materials in aerospace industry, Integrated Ferroelectrics 211 (1) (2020) 25–44.

[10] Hassan Elahi, Marco Eugeni, Paolo Gaudenzi, Madiha Gul, Raees Fida Swati, Piezoelectric thermo electromechanical energy harvester for reconnaissance satellite structure, Microsystem Technologies 25 (2) (2019) 665–672.

[11] Hassan Elahi, Marco Eugeni, Paolo Gaudenzi, Faisal Qayyum, Raees Fida Swati, Hayat Muhammad Khan, Response of piezoelectric materials on thermomechanical shocking and electrical shocking for aerospace applications, Microsystem Technologies 24 (9) (2018) 3791–3798.

[12] Hassan Elahi, Zubair Butt, Marco Eugnei, Paolo Gaudenzi, Asif Israr, Effects of variable resistance on smart structures of cubic reconnaissance satellites in various thermal and frequency shocking conditions, Journal of Mechanical Science and Technology 31 (9) (2017) 4151–4157.

[13] Mohsin Ali Marwat, Weigang Ma, Pengyuan Fan, Hassan Elahi, Chanatip Samart, Bo Nan, Hua Tan, David Salamon, Baohua Ye, Haibo Zhang, Ultrahigh energy density and thermal stability in sandwich-structured nanocomposites with dopamine@Ag@BaTiO$_3$, Energy Storage Materials 31 (2020) 492–504.

[14] Ahsan Ali, Riffat Asim Pasha, Hassan Elahi, Muhammad Abdullah Sheeraz, Saima Bibi, Zain Ul Hassan, Marco Eugeni, Paolo Gaudenzi, Investigation of deformation in bimorph piezoelectric actuator: analytical, numerical and experimental approach, Integrated Ferroelectrics 201 (1) (2019) 94–109.

[15] Muhammad Usman Khan, Zubair Butt, Hassan Elahi, Waqas Asghar, Zulkarnain Abbas, Muhammad Shoaib, M. Anser Bashir, Deflection of coupled elasticity–electrostatic bimorph PVDF material: theoretical, FEM and experimental verification, Microsystem Technologies 25 (8) (2019) 3235–3242.

[16] H. Elahi, A. Israr, R.F. Swati, H.M. Khan, A. Tamoor, Stability of piezoelectric material for suspension applications, in: 2017 Fifth International Conference on Aerospace Science & Engineering (ICASE), IEEE, 2017, pp. 1–5.

[17] Mohsen Safaei, Henry A. Sodano, Steven R. Anton, A review of energy harvesting using piezoelectric materials: state-of-the-art a decade later (2008–2018), Smart Materials and Structures 28 (11) (2019) 113001.

[18] S.R. Anton, A. Erturk, D.J. Inman, Piezoelectric energy harvesting from multifunctional wing spars for UAVs: Part 2. Experiments and storage applications, in: Active and Passive Smart Structures and Integrated Systems 2009, vol. 7288, International Society for Optics and Photonics, 2009, 72880D.

[19] Ahmadreza Tabesh, Luc G. Fréchette, A low-power stand-alone adaptive circuit for harvesting energy from a piezoelectric micropower generator, IEEE Transactions on Industrial Electronics 57 (3) (2009) 840–849.

[20] Elie Lefeuvre, David Audigier, Claude Richard, Daniel Guyomar, Buck-boost converter for sensorless power optimization of piezoelectric energy harvester, IEEE Transactions on Power Electronics 22 (5) (2007) 2018–2025.

[21] Elie Lefeuvre, Adrien Badel, Claude Richard, Daniel Guyomar, Piezoelectric energy harvesting device optimization by synchronous electric charge extraction, Journal of Intelligent Material Systems and Structures 16 (10) (2005) 865–876.

[22] Daniel Guyomar, Adrien Badel, Elie Lefeuvre, Claude Richard, Toward energy harvesting using active materials and conversion improvement by nonlinear processing, IEEE Transactions on Ultrasonics, Ferroelectrics, and Frequency Control 52 (4) (2005) 584–595.

[23] Adrien Badel, Daniel Guyomar, Elie Lefeuvre, Claude Richard, Efficiency enhancement of a piezoelectric energy harvesting device in pulsed operation by synchronous charge inversion, Journal of Intelligent Material Systems and Structures 16 (10) (2005) 889–901.

[24] Elie Lefeuvre, Adrien Badel, Claude Richard, Lionel Petit, Daniel Guyomar, A comparison between several vibration-powered piezoelectric generators for stand-alone systems, Sensors and Actuators A: Physical 126 (2) (2006) 405–416.

[25] Mickaël Lallart, Lauric Garbuio, Lionel Petit, Claude Richard, Daniel Guyomar, Double synchronized switch harvesting (DSSH): a new energy harvesting scheme for efficient energy extraction, IEEE Transactions on Ultrasonics, Ferroelectrics, and Frequency Control 55 (10) (2008) 2119–2130.

[26] Paul C-P. Chao, Energy harvesting electronics for vibratory devices in self-powered sensors, IEEE Sensors Journal 11 (12) (2011) 3106–3121.

[27] Hui Shen, Jinhao Qiu, Hongli Ji, Kongjun Zhu, Marco Balsi, Enhanced synchronized switch harvesting: a new energy harvesting scheme for efficient energy extraction, Smart Materials and Structures 19 (11) (2010) 115017.

[28] Lauric Garbuio, Mickaël Lallart, Daniel Guyomar, Claude Richard, David Audigier, Mechanical energy harvester with ultralow threshold rectification based on SSHI nonlinear technique, IEEE Transactions on Industrial Electronics 56 (4) (2009) 1048–1056.

[29] Mickaël Lallart, Claude Richard, Lauric Garbuio, Lionel Petit, Daniel Guyomar, High efficiency, wide load bandwidth piezoelectric energy scavenging by a hybrid nonlinear approach, Sensors and Actuators A: Physical 165 (2) (2011) 294–302.

[30] Junrui Liang, Wei-Hsin Liao, Improved design and analysis of self-powered synchronized switch interface circuit for piezoelectric energy harvesting systems, IEEE Transactions on Industrial Electronics 59 (4) (2011) 1950–1960.

[31] Adrien Badel, Elie Lefeuvre, Nonlinear conditioning circuits for piezoelectric energy harvesters, in: Nonlinearity in Energy Harvesting Systems, Springer, 2016, pp. 321–359.

[32] Alexis Brenes, Elie Lefeuvre, Adrien Badel, Seonho Seok, Chan-Sei Yoo, Unipolar synchronized electric charge extraction for piezoelectric energy harvesting, Smart Materials and Structures 27 (7) (2018) 075054.

[33] Kang Zhao, Junrui Liang, Haoyu Wang, Series synchronized triple bias-flip (S-S3BF) interface circuit for piezoelectric energy harvesting, in: 2019 IEEE International Symposium on Circuits and Systems (ISCAS), IEEE, 2019, pp. 1–5.

[34] Herbert W. Jackson, Introduction to Electric Circuits, 1970.

[35] Hassan Elahi, Khushboo Munir, Marco Eugeni, Paolo Gaudenzi, Reliability risk analysis for the aeroelastic piezoelectric energy harvesters, Integrated Ferroelectrics 212 (1) (2020) 156–169.

[36] Hassan Elahi, Marco Eugeni, Luca Lampani, Paolo Gaudenzi, Modeling and design of a piezoelectric nonlinear aeroelastic energy harvester, Integrated Ferroelectrics 211 (1) (2020) 132–151.

[37] Hassan Elahi, Marco Eugeni, Federico Fune, Luca Lampani, Franco Mastroddi, Giovanni Paolo Romano, Paolo Gaudenzi, Performance evaluation of a piezoelectric energy harvester based on flag-flutter, Micromachines 11 (10) (2020) 933.

[38] Marco Eugeni, Hassan Elahi, Federico Fune, Luca Lampani, Franco Mastroddi, Giovanni Paolo Romano, Paolo Gaudenzi, Numerical and experimental investigation of piezoelectric energy harvester based on flag-flutter, Aerospace Science and Technology 97 (2020) 105634.

[39] Hassan Elahi, Marco Eugeni, Paolo Gaudenzi, Design and performance evaluation of a piezoelectric aeroelastic energy harvester based on the limit cycle oscillation phenomenon, Acta Astronautica 157 (2019) 233–240.

[40] Marco Eugeni, Hassan Elahi, Federico Fune, Luca Lampani, Franco Mastroddi, Giovanni Paolo Romano, Paolo Gaudenzi, Experimental evaluation of piezoelectric energy harvester based on flag-flutter, in: Conference of the Italian Association of Theoretical and Applied Mechanics, Springer, 2019, pp. 807–816.

[41] N.A. Kong, Dong Sam Ha, Alper Erturk, Daniel J. Inman, Resistive impedance matching circuit for piezoelectric energy harvesting, Journal of Intelligent Material Systems and Structures 21 (13) (2010) 1293–1302.

[42] Hyeoungwoo Kim, Shashank Priya, Harry Stephanou, Kenji Uchino, Consideration of impedance matching techniques for efficient piezoelectric energy harvesting, IEEE Transactions on Ultrasonics, Ferroelectrics, and Frequency Control 54 (9) (2007) 1851–1859.

[43] Daniel Guyomar, Mickaël Lallart, Recent progress in piezoelectric conversion and energy harvesting using nonlinear electronic interfaces and issues in small scale implementation, Micromachines 2 (2) (2011) 274–294.

[44] A. Brenes, E. Lefeuvre, C-S. Yoo, Experimental validation of wideband piezoelectric energy harvesting based on frequency-tuning synchronized charge extraction, Journal of Physics: Conference Series 1052 (2018) 012050.

[45] Alexis Brenes, Elie Lefeuvre, Adrien Badel, Seonho Seok, Chan-Sei Yoo, Shunt-diode rectifier: a new scheme for efficient piezoelectric energy harvesting, Smart Materials and Structures 28 (1) (2018) 015015.

[46] Björn Richter, Jens Twiefel, Jörg Wallaschek, Piezoelectric equivalent circuit models, in: Energy Harvesting Technologies, Springer, 2009, pp. 107–128.

[47] Hassan Elahi, Marco Eugeni, Paolo Gaudenzi, Electromechanical degradation of piezoelectric patches, in: Analysis and Modelling of Advanced Structures and Smart Systems, Springer, 2018, pp. 35–44.

[48] Vittorio Memmolo, Hassan Elahi, Marco Eugeni, Ernesto Monaco, Fabrizio Ricci, Michele Pasquali, Paolo Gaudenzi, Experimental and numerical investigation of PZT response in composite structures with variable degradation levels, Journal of Materials Engineering and Performance 28 (6) (2019) 3239–3246.

[49] Stefano Saggini, Stefano Giro, Fabio Ongaro, Paolo Mattavelli, Implementation of reactive and resistive load matching for optimal energy harvesting from piezoelectric generators, in: 2010 IEEE 12th Workshop on Control and Modeling for Power Electronics (COMPEL), IEEE, 2010, pp. 1–6.

[50] Na Kong, Dong Sam Ha, Low-power design of a self-powered piezoelectric energy harvesting system with maximum power point tracking, IEEE Transactions on Power Electronics 27 (5) (2011) 2298–2308.

[51] J. Sankman, D. Ma, A 12-μW to 1.1-mW AIM piezoelectric energy harvester for time-varying vibrations with 450-nA, IEEE Transactions on Power Electronics 30 (2) (2015) 632–643.

[52] Chao Lu, Chi-Ying Tsui, Wing-Hung Ki, Vibration energy scavenging system with maximum power tracking for micropower applications, IEEE Transactions on Very Large Scale Integration (VLSI) Systems 19 (11) (2010) 2109–2119.

[53] K.R. Bharath, Eenisha Suresh, Design and implementation of improved fractional open circuit voltage based maximum power point tracking algorithm for photovoltaic applications, International Journal of Renewable Energy Research (IJRER) 7 (3) (2017) 1108–1113.

[54] Yogesh K. Ramadass, Anantha P. Chandrakasan, An efficient piezoelectric energy harvesting interface circuit using a bias-flip rectifier and shared inductor, IEEE Journal of Solid-State Circuits 45 (1) (2009) 189–204.

[55] Liao Wu, Xuan-Dien Do, Sang-Gug Lee, Dong Sam Ha, A self-powered and optimal SSHI circuit integrated with an active rectifier for piezoelectric energy harvesting, IEEE Transactions on Circuits and Systems I: Regular Papers 64 (3) (2016) 537–549.

[56] S. Saeid Hashemi, Mohamad Sawan, Yvon Savaria, A high-efficiency low-voltage CMOS rectifier for harvesting energy in implantable devices, IEEE Transactions on Biomedical Circuits and Systems 6 (4) (2012) 326–335.

[57] Yat-Hei Lam, Wing-Hung Ki, Chi-Ying Tsui, Integrated low-loss CMOS active rectifier for wirelessly powered devices, IEEE Transactions on Circuits and Systems II: Express Briefs 53 (12) (2006) 1378–1382.

[58] Triet T. Le, Jifeng Han, Annette Von Jouanne, Kartikeya Mayaram, Terri S. Fiez, Piezoelectric micro-power generation interface circuits, IEEE Journal of Solid-State Circuits 41 (6) (2006) 1411–1420.

[59] Shashank Priya, Hyun-Cheol Song, Yuan Zhou, Ronnie Varghese, Anuj Chopra, Sang-Gook Kim, Isaku Kanno, Liao Wu, Dong Sam Ha, Jungho Ryu, et al., A review on piezoelectric energy harvesting: materials, methods, and circuits, Energy Harvesting and Systems 4 (1) (2019) 3–39.

[60] Geon-Tae Hwang, Joonseok Yang, Seong Ho Yang, Ho-Yong Lee, Minbok Lee, Dae Yong Park, Jae Hyun Han, Seung Jun Lee, Chang Kyu Jeong, Jaeha Kim, et al., A reconfigurable rectified flexible energy harvester via solid-state single crystal grown PMN–PZT, Advanced Energy Materials 5 (10) (2015) 1500051.

[61] Gabriel A. Rincón-Mora, Siyu Yang, Tiny piezoelectric harvesters: principles, constraints, and power conversion, IEEE Transactions on Circuits and Systems I: Regular Papers 63 (5) (2016) 639–649.

CHAPTER 6

Modeling and simulation of a piezoelectric energy harvester

Contents

6.1.	Introduction	99
6.2.	Modeling	100
6.3.	Material creation	103
6.4.	Material assignment	104
6.5.	Interaction	107
6.6.	Input step creation	108
6.7.	Output assignment	110
6.8.	Boundary conditions	112
6.9.	Loading conditions	112
6.10.	Meshing	113
6.11.	Job creation	117
6.12.	Results	117
References		120

6.1 Introduction

Modeling simulation resolves challenges in the physical world safely and effectively. It offers a valuable and easy to validate, contact, and understand form of study. Simulation offers useful solutions through simple insights into dynamic processes spanning sectors and disciplines [1,2]. Simulation makes experiments with a true digital device representation. In comparison to physical modelings, such as copying a building in size, simulation modeling is based on computers and uses algorithms and equations [3,4]. In order to evaluate machine models in operation, simulation software creates a dynamic world with the option to display in 2D or 3D [5–7].

In this chapter, modeling and simulation of the piezoelectric energy harvester are explained in detail. Several commercial finite element sources are available that includes the modeling of piezoelectric behavior such as ADINA. Abaqus© is a highly advanced, general-purpose finite element software, primarily aimed at modeling solid and structural behavior under external loading [8–10]. Piezoelectric materials are playing a vital role in energy harvesting processes [11–13]. This chapter includes geometric modeling capabilities with a comprehensive element library for static and

Piezoelectric Aeroelastic Energy Harvesting
https://doi.org/10.1016/B978-0-12-823968-1.00017-9

Copyright © 2022 Elsevier Inc.
All rights reserved.

99

dynamic studies using linear and non-linear constitutive relations in small and large deformation processes. It has an up-to-date and well explained mechanical and computational theoretical history and is commonly used in industry and academic research work.

Energy harvesting via piezoelectric transduction is one of the most promising tasks in the aerospace industry [14–17]. The overall campaign carried out during the modeling and simulation in Abaqus is presented in Fig. 6.1. These steps are constituted of modeling, material creation, material assignment, interaction, boundary conditions, loading conditions, meshing, output assignment, job creation, and results. All these steps are explained in detail with screenshots of the Abaqus software so that readers can understand in a better way. The basic steps are almost the same as in any commercially available software for structural analysis, these concepts can be applied to other software accordingly.

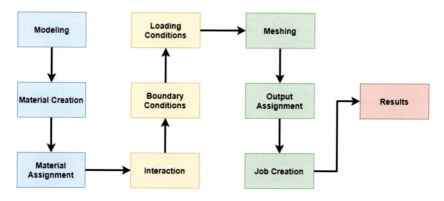

Figure 6.1 Common steps carried out during modeling and simulation of a piezoelectric energy harvester.

6.2 Modeling

The first step carried out in any commercially available simulation software, e.g., Abaqus, is to draw the model as per engineering design or to import the model from any CAD software. In this subsection, the reader will understand how to model a structure in Abaqus. As discussed earlier, our aim is to design a piezoelectric energy harvester.

In Fig. 6.2, the home screen of Abaqus software is shown. To create a model as per the engineering design, double click on "Model-1". Fig. 6.3 will pop-up on the screen presenting input parameters for the model creation. In this tab, we have to decide on the modeling space, type, base feature, and approximate size as per engineering design. For a piezoelectric energy harvester, 3D, deformable, solid extrusion, and 200 approximation size is chosen.

Figure 6.2 Abaqus software home screen.

Figure 6.3 Model creation input parameters.

After selecting the approximate size, the grid size appears on the Abaqus home screen as shown in Fig. 6.4. This grid will help model the parts and is easy to understand the dimensions.

In Fig. 6.5, the geometry of the piezoelectric harvester is represented. The important point is that Abaqus is unitless, for the model development use the same standards of the unit. The dimensions of the harvester are represented in Figs. 6.5 and 6.6. The model extrusion is presented in Fig. 6.6 for 3D modeling. The 3D model of the harvester is presented in Fig. 6.7.

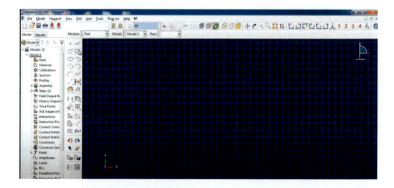

Figure 6.4 Grid size for model creation.

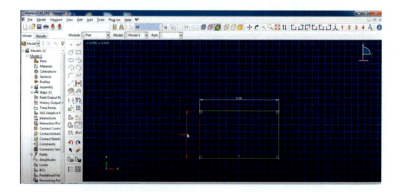

Figure 6.5 Geometry of the piezoelectric patch.

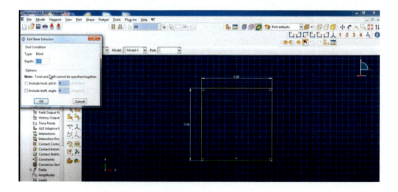

Figure 6.6 Extrusion of the model.

Modeling and simulation of a piezoelectric energy harvester 103

Figure 6.7 3D model of an energy harvester.

6.3 Material creation

In this phase, material properties are created for the model that has been created in the previous step. The number of materials created depends on the number of different materials that are used in the model. For the current model, i.e., piezoelectric energy harvester, the model only consists of piezoelectric material. Therefore, only piezoelectric material is created in this phase.

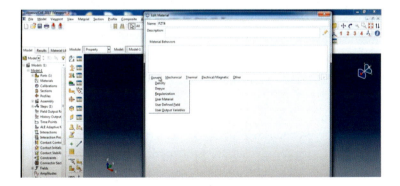

Figure 6.8 Material creation.

For material creation, double click on "Material" and the "Edit Material" tab will pop-up as shown in Fig. 6.8. From this tab, the density of the material is assigned as shown in Fig. 6.9. These properties are assigned as per material characteristics. The elastic properties of piezoelectric materials

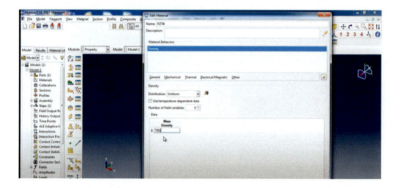

Figure 6.9 Assigning density to piezoelectric material.

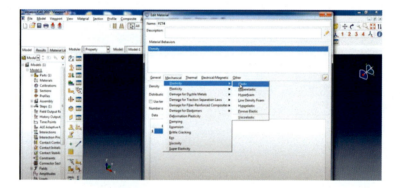

Figure 6.10 Assigning elastic properties.

are defined as shown in Fig. 6.10. The material properties can also be seen in [18–21].

The orthotropic properties are illustrated in Fig. 6.11. In this tab, strain properties of the piezoelectric patch are assigned as shown in Fig. 6.12. Anisotropic and stress properties are presented in Figs. 6.13 and 6.14, respectively.

6.4 Material assignment

In this phase, the constituted materials (that were created in the previous step) are assigned to the model. Therefore, the model can behave as designed for the given conditions. Fig. 6.15 represents the section creation for the harvester. The material that was created in the previous step was

Modeling and simulation of a piezoelectric energy harvester 105

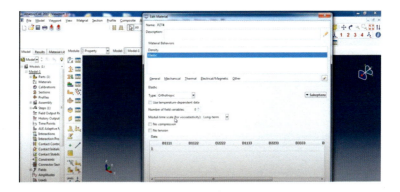

Figure 6.11 Orthotropic properties assignment.

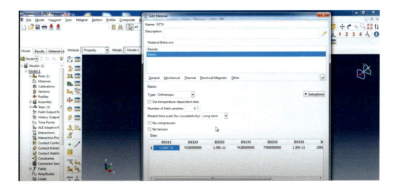

Figure 6.12 Assigning strain values.

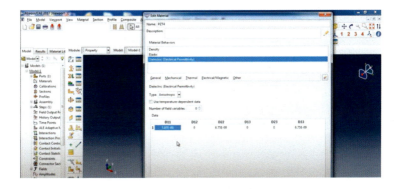

Figure 6.13 Anisotropic values.

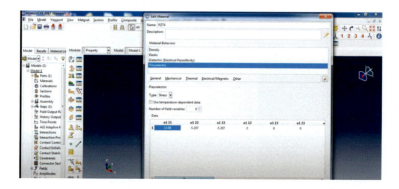

Figure 6.14 Assigning stress values.

Figure 6.15 Create section.

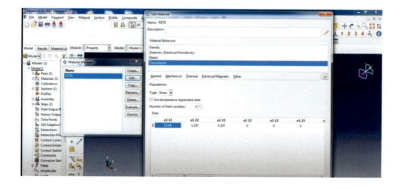

Figure 6.16 Material assignment.

Modeling and simulation of a piezoelectric energy harvester 107

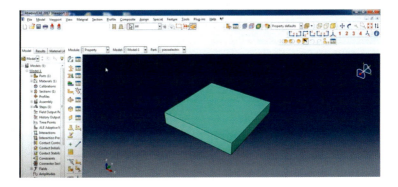

Figure 6.17 Material assigned.

assigned to the model in this step as shown in Fig. 6.16. After assigning the material to the model, the color of the model is converted from white as shown in Fig. 6.17.

6.5 Interaction

In this phase, the interactions in the model parts are created. The interaction of the various parts is defined. In complex geometries, the various links and parts are dependent on the motion of other links or parts. Such restricted motions are defined here.

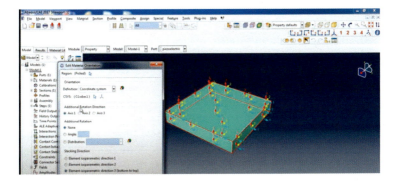

Figure 6.18 Editing material orientation.

The material orientation and degrees assigned to the rotation are presented in Figs. 6.18 and 6.19. The assembly of the model is defined in

Figure 6.19 Material orientation rotation.

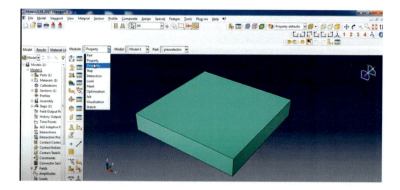

Figure 6.20 Assembly.

Fig. 6.20 and the instances for the piezoelectric harvester are defined in Fig. 6.21.

6.6 Input step creation

In this phase, the type of loading conditions, i.e., the input conditions are defined. These conditions are defined according to the real experimentation data or by following considerations of the engineering design. The input steps are defined in Fig. 6.22. Even the numerous input steps can also be defined in this phase as per the conditions. Fig. 6.23 presents the time period for the input step. The increment size for the input step time period is shown in Fig. 6.24. The different options for step input as per the engineering design can be seen in Fig. 6.25.

Modeling and simulation of a piezoelectric energy harvester 109

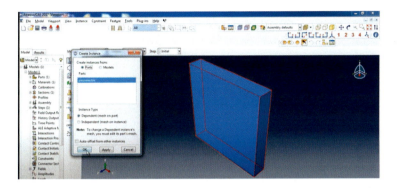

Figure 6.21 Create instances.

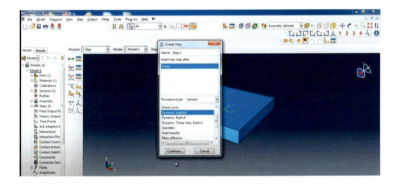

Figure 6.22 Input step for loading condition.

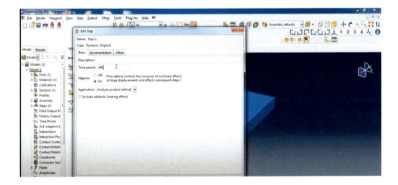

Figure 6.23 Time period of the input step.

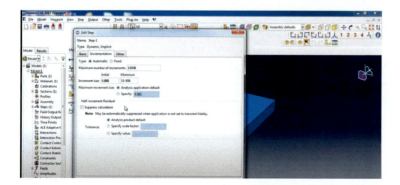

Figure 6.24 Increment size for the input step time period.

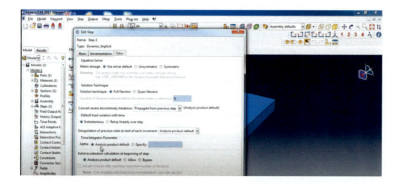

Figure 6.25 Different options for step input.

6.7 Output assignment

In this phase, users can define the parameters which they require in the output of the analysis. What type of response do we want to analyze for the given input? Such details can be analyzed in this step. For example, as a result of applied load, we are interested in seeing the overall stress and strain of the system, then we have to select stress and strain from "Create Output."

In this case, we are interested in analyzing the electric potential generated by the piezoelectric energy harvester. Therefore, we selected "EPOT" from the output tab. In Fig. 6.26, the field output manager tab is shown. The output electrical potential is selected in the field output manager as represented in Fig. 6.27. The output is then associated with the input step created as represented in Fig. 6.28.

Modeling and simulation of a piezoelectric energy harvester   111

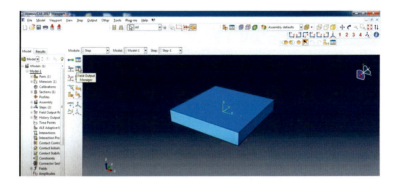

Figure 6.26 Field output manager.

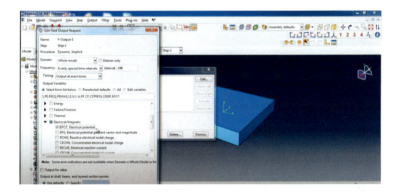

Figure 6.27 Electrical potential output.

Figure 6.28 Output creation for input step.

6.8 Boundary conditions

In this phase, the boundary conditions of the model are defined. For example, if we have to analyze the cantilever beam then one end of the beam is considered to be rigid, which is done in this phase. For the case of the piezoelectric energy harvester, the boundary condition creation is represented in Fig. 6.29, electrical boundary conditions are presented in Fig. 6.30, and overall boundary conditions are shown in Fig. 6.31.

Figure 6.29 Boundary conditions creation.

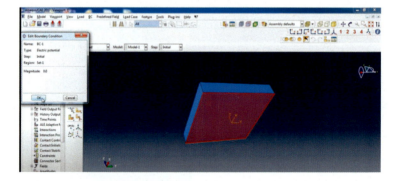

Figure 6.30 Electric potential boundary condition.

6.9 Loading conditions

In this phase, the loading conditions are created based on the input steps created in the previous step. At this stage, loading conditions are created

Modeling and simulation of a piezoelectric energy harvester   113

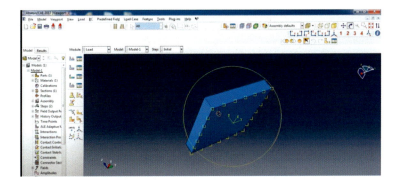

Figure 6.31 Overall boundary conditions of the model.

as represented in Fig. 6.32. After creating the load, the parameters are defined in this step, i.e., magnitude, distribution, and amplitude of load, as represented in Fig. 6.33. For the selected the parametric values, the overall loading conditions are represented in Fig. 6.34. This representation makes it easy to understand the physical terms of loading conditions.

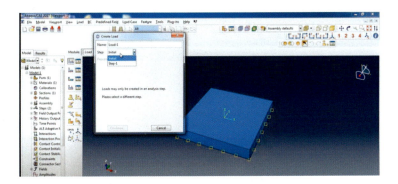

Figure 6.32 Load creation for a piezoelectric energy harvester.

6.10 Meshing

Meshing is an important part of the engineering simulation process, where complex geometries are divided into simple components that are ideal for using a discerning local approach to a wide area. The mesh affects the precision, convergence, and speed of the simulation.

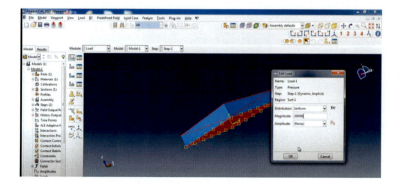

Figure 6.33 Loading parameters.

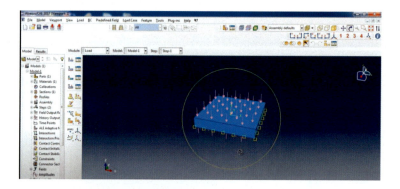

Figure 6.34 Overall load representation.

The most popular meshing method used in Abaqus/CAE is done top-down. This implies that the mesh is rendered exactly according to the geometry of a region and works to the location of the element and node. There are many advantages and drawbacks of using the standardized mesh in any aspect of grid development. The benefit of such a mesh is that a double index (i, j) or a three-dimensional index (i, j, k) can easily address the points in an elemental cell. Connectivity is simple because the indexes are used to classify cells adjacent to a given elemental face, and the cell's edges form continuous mesh lines that begin on opposite elemental sides. In 2D, four adjacent cells bind the central cell. Six surrounding cells are connected to the central cell in three dimensions. The structured mesh also enables quick data management and ongoing networking, making it easy to program. Non-orthogonal or skewing grid growth, which can cause unphysical solutions due to the transformation of the governing equations, is nonetheless an inconvenience for such a mesh, particularly for more complex geometries. The transformed equations, which are suitable for non-orthogonal grids, connect the standardized system of coordinates (such as Cartesian coordinates) and the body-built system of coordinates with supplementary terms and thus increase both the cost of numerical calculations and program complexity. This can also impact the exactness and efficacy of the implemented numerical algorithm. In CFD applications, the use of an unstructured mesh is growing. Today, the majority of industrial codes are based on the unstructured mesh approach. The cells can be freely installed inside the machine domain here. The connectivity information for each side, therefore, needs proper storage in a table shape. A triangle of two dimensions or a tetrahedron of three dimensions is the most typical shape of an unstructured element. Any other basic form, like quadrilateral or hexahedral cell, is nevertheless possible.

The reader should also know, considering its various benefits, the drawbacks of using a non-structured mesh for CFD simulation. In relation to the structured mesh, the elementary cell points for an unstructured mesh may typically only be handled or discussed by two-dimensional double (i, j) or three-dimensional tripple (i, j, k) indices. An elementary cell can be connected to an infinite number of neighboring cells, which makes data processing and communication difficult.

Triangular (two-dimensional) or tetrahedral (three-dimensional) cells are typically ineffective in the resolution of wall boundary layers in contrast with a quadrilateral (two-dimensional) and hexahedral (three-dimensional) cells. In most cases, the grid yields very long, thin, triangular, or tetrahedral cells adjacent to the wall limits and thus leads to major diffusive flux approximations. The need for a more complex solution algorithm in order to resolve flow-field variables is another drawback in combination with data processing and connectivity of elementary cells. This could lead to longer machine time when a solution is obtained and the improvements in computer performance may become worse.

For the case of piezoelectric energy harvester, the global and local meshing are represented in Figs. 6.35 and 6.36, respectively. The overall meshing of the piezoelectric energy harvester is presented in Fig. 6.37. The element type for meshing is represented in Fig. 6.38. The element type is chosen to be piezoelectric for energy harvesting purposes.

116 Piezoelectric Aeroelastic Energy Harvesting

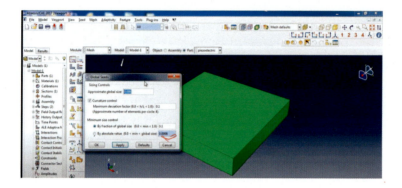

Figure 6.35 Global seeds for meshing.

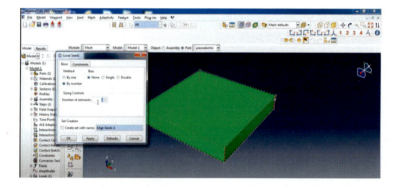

Figure 6.36 Local seeds for meshing.

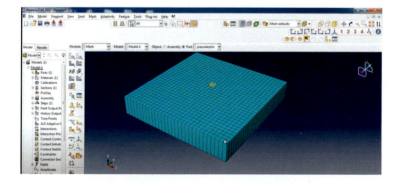

Figure 6.37 Overall meshing of the piezoelectric energy harvester.

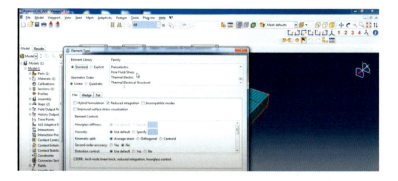

Figure 6.38 Element type for meshing.

6.11 Job creation

Double click the "Jobs" tab in the "Model Tree" to construct a job of the model completed and save it with the desired name. The "Work" module is moved to Abaqus/CAE and the "Generates Job" dialog box appears in the model database with a list of models. Expand your "Jobs" container in the model tree; click on the job that was created, and pick "Send" from your job analysis menu. After you apply for the job, information on the status of the job appears next to the name of the job. One of the following is shown as the status of the piezoelectric energy harvester: "Submitted" during input analyses; "Run" during the study of a model by Abaqus; "Completed" when the analysis is complete and the output to the output database is written; and "Aborted" if the input or analysis file problem is detected by the Abaqus/CAE and the analysis is aborted. Moreover, in the message field, Abaqus/CAE reports the issue. The job creation for the piezoelectric energy harvester is represented in Fig. 6.39.

6.12 Results

When the job is submitted successfully, the results are analyzed in Abaqus. These results help the user understand the physics behind the designed model in a better way. The overall stresses calculated by the given loading conditions are presented in Fig. 6.40. The electric potential that can be achieved from the harvester is represented in Fig. 6.41. In order to plot the data in XY coordinates or in tabular form, double click "XY Data" on "Output Field" and select the desired parameters as represented in Figs. 6.42, 6.43, and 6.44.

118 Piezoelectric Aeroelastic Energy Harvesting

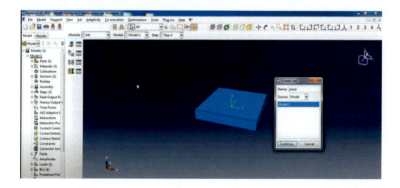

Figure 6.39 Job creation.

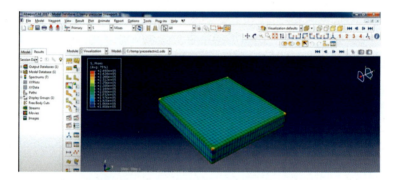

Figure 6.40 Stresses calculation of the piezoelectric energy harvester.

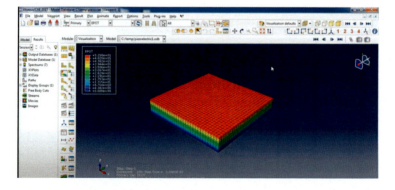

Figure 6.41 Electric potential generated by the piezoelectric energy harvester.

Modeling and simulation of a piezoelectric energy harvester 119

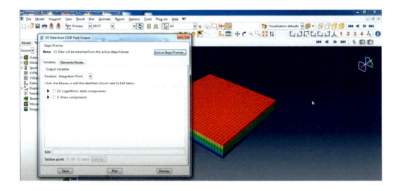

Figure 6.42 XY plot creation.

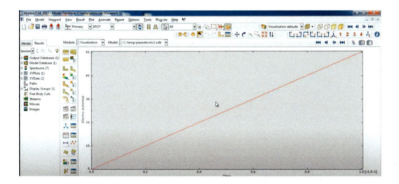

Figure 6.43 Electric potential vs time in graphical format.

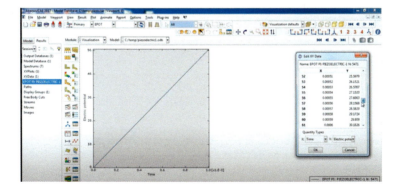

Figure 6.44 Electric potential vs time in tabular format.

References

[1] Hassan Elahi, Khushboo Munir, Marco Eugeni, Paolo Gaudenzi, Reliability risk analysis for the aeroelastic piezoelectric energy harvesters, Integrated Ferroelectrics 212 (1) (2020) 156–169.

[2] Hassan Elahi, Marco Eugeni, Luca Lampani, Paolo Gaudenzi, Modeling and design of a piezoelectric nonlinear aeroelastic energy harvester, Integrated Ferroelectrics 211 (1) (2020) 132–151.

[3] Hassan Elahi, Marco Eugeni, Paolo Gaudenzi, Design and performance evaluation of a piezoelectric aeroelastic energy harvester based on the limit cycle oscillation phenomenon, Acta Astronautica 157 (2019) 233–240.

[4] Marco Eugeni, Hassan Elahi, Federico Fune, Luca Lampani, Franco Mastroddi, Giovanni Paolo Romano, Paolo Gaudenzi, Experimental evaluation of piezoelectric energy harvester based on flag-flutter, in: Conference of the Italian Association of Theoretical and Applied Mechanics, Springer, 2019, pp. 807–816.

[5] Hassan Elahi, Marco Eugeni, Federico Fune, Luca Lampani, Franco Mastroddi, Giovanni Paolo Romano, Paolo Gaudenzi, Performance evaluation of a piezoelectric energy harvester based on flag-flutter, Micromachines 11 (10) (2020) 933.

[6] Hassan Elahi, The investigation on structural health monitoring of aerospace structures via piezoelectric aeroelastic energy harvesting, Microsystem Technologies (2020) 1–9.

[7] Marco Eugeni, Hassan Elahi, Federico Fune, Luca Lampani, Franco Mastroddi, Giovanni Paolo Romano, Paolo Gaudenzi, Numerical and experimental investigation of piezoelectric energy harvester based on flag-flutter, Aerospace Science and Technology 97 (2020) 105634.

[8] Hassan Elahi, Marco Eugeni, Paolo Gaudenzi, Electromechanical degradation of piezoelectric patches, in: Analysis and Modelling of Advanced Structures and Smart Systems, Springer, 2018, pp. 35–44.

[9] H. Elahi, A. Israr, R.F. Swati, H.M. Khan, A. Tamoor, Stability of piezoelectric material for suspension applications, in: 2017 Fifth International Conference on Aerospace Science & Engineering (ICASE), IEEE, 2017, pp. 1–5.

[10] Vittorio Memmolo, Hassan Elahi, Marco Eugeni, Ernesto Monaco, Fabrizio Ricci, Michele Pasquali, Paolo Gaudenzi, Experimental and numerical investigation of PZT response in composite structures with variable degradation levels, Journal of Materials Engineering and Performance 28 (6) (2019) 3239–3246.

[11] Muhammad Usman Khan, Zubair Butt, Hassan Elahi, Waqas Asghar, Zulkarnain Abbas, Muhammad Shoaib, M. Anser Bashir, Deflection of coupled elasticity–electrostatic bimorph PVDF material: theoretical, FEM and experimental verification, Microsystem Technologies 25 (8) (2019) 3235–3242.

[12] Ahsan Ali, Riffat Asim Pasha, Hassan Elahi, Muhammad Abdullah Sheeraz, Saima Bibi, Zain Ul Hassan, Marco Eugeni, Paolo Gaudenzi, Investigation of deformation in bimorph piezoelectric actuator: analytical, numerical and experimental approach, Integrated Ferroelectrics 201 (1) (2019) 94–109.

[13] Mohsin Ali Marwat, Weigang Ma, Pengyuan Fan, Hassan Elahi, Chanatip Samart, Bo Nan, Hua Tan, David Salamon, Baohua Ye, Haibo Zhang, Ultrahigh energy density and thermal stability in sandwich-structured nanocomposites with dopamine@Ag@BaTiO$_3$, Energy Storage Materials 31 (2020) 492–504.

[14] Hassan Elahi, Khushboo Munir, Marco Eugeni, Muneeb Abrar, Asif Khan, Adeel Arshad, Paolo Gaudenzi, A review on applications of piezoelectric materials in aerospace industry, Integrated Ferroelectrics 211 (1) (2020) 25–44.

[15] Hassan Elahi, Marco Eugeni, Paolo Gaudenzi, Madiha Gul, Raees Fida Swati, Piezoelectric thermo electromechanical energy harvester for reconnaissance satellite structure, Microsystem Technologies 25 (2) (2019) 665–672.

[16] Hassan Elahi, Marco Eugeni, Paolo Gaudenzi, Faisal Qayyum, Raees Fida Swati, Hayat Muhammad Khan, Response of piezoelectric materials on thermomechanical shocking and electrical shocking for aerospace applications, Microsystem Technologies 24 (9) (2018) 3791–3798.

[17] Hassan Elahi, Zubair Butt, Marco Eugnei, Paolo Gaudenzi, Asif Israr, Effects of variable resistance on smart structures of cubic reconnaissance satellites in various thermal and frequency shocking conditions, Journal of Mechanical Science and Technology 31 (9) (2017) 4151–4157.

[18] Hassan Elahi, M. Rizwan Mughal, Marco Eugeni, Faisal Qayyum, Asif Israr, Ahsan Ali, Khushboo Munir, Jaan Praks, Paolo Gaudenzi, Characterization and implementation of a piezoelectric energy harvester configuration: analytical, numerical and experimental approach, Integrated Ferroelectrics 212 (1) (2020) 39–60.

[19] Hassan Elahi, Khushboo Munir, Marco Eugeni, Sofiane Atek, Paolo Gaudenzi, Energy harvesting towards self-powered IoT devices, Energies 13 (21) (2020) 5528.

[20] Hassan Elahi, Marco Eugeni, Paolo Gaudenzi, A review on mechanisms for piezoelectric-based energy harvesters, Energies 11 (7) (2018) 1850.

[21] Zubair Butt, Riffat Asim Pasha, Faisal Qayyum, Zeeshan Anjum, Nasir Ahmad, Hassan Elahi, Generation of electrical energy using lead zirconate titanate (PZT-5a) piezoelectric material: analytical, numerical and experimental verifications, Journal of Mechanical Science and Technology 30 (8) (2016) 3553–3558.

PART 3

Aeroelastic energy harvesting

CHAPTER 7

Fluid–structure interaction: some issues about the aeroelastic problem

Contents

7.1.	Introduction	125
	7.1.1 Basic definitions of stability	126
7.2.	Bifurcation problems	127
	7.2.1 Hopf bifurcation	128
7.3.	Aeroelastic problem formulation	128
	7.3.1 p–k method	128
	7.3.1.1 Iterative method	*129*
7.4.	Finite element method for flag-flutter	130
	7.4.1 Numerical model	130
7.5.	2D modeling for steady FSI system	131
	7.5.1 2D structures	131
	7.5.2 2D aerodynamic surfaces	132
	7.5.3 FSI system	133
7.6.	Dynamic FSI	133
7.7.	Aerodynamic theories	136
	7.7.1 Wagner function	137
	7.7.2 Kussner function	138
7.8.	General approximation	139
	7.8.1 Strip theory	139
	7.8.2 Quasi-steady	139
	7.8.3 Slender body/wing	140
References		141

7.1 Introduction

In a system where the interaction between fluid flow and a structure is considered, it is important to monitor the overall fluid dynamics built by the whole system; this type of system is named an aeroelastic system [1]. This chapter discusses the issues encountered by such aeroelastic systems. To have a full comprehension of the aeroelastic problem, it is compulsory to understand what aeroelasticity actually means. It is a field that studies the mutual interaction of an elastic structure deformation with a surrounding

---

Piezoelectric Aeroelastic Energy Harvesting
https://doi.org/10.1016/B978-0-12-823968-1.00012-X

Copyright © 2022 Elsevier Inc.
All rights reserved.

125

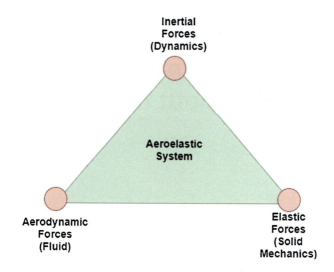

Figure 7.1 Collar's triangle for aeroelasticity.

flow and the corresponding aerodynamic forces [2]. Observing the Collar's triangle, we come to know the multidisciplinary nature of aeroelasticity as shown in Fig. 7.1. Generally, there are two types of phenomenon: the first is the *static aeroelastic* phenomenon, which is outside the triangle, and the second is the *dynamic aeroelastic* phenomenon, which lies within the triangle, involving all the forces. They have numerous applications in the field of aerospace engineering [3–6].

The problem arises from the mutual interaction between an elastic structure and a flow. It is basically governed by a variety of parameters that determine flow conditions. Some of them are the unperturbed speed or the Mach number. These parameters influence the system solution stability and are responsible for bringing the equilibrium to any unstable condition [7,8]. In this chapter some basic issues of aeroelasticity are addressed in order to prepare the discussion concerning the Energy Harversting applications of FSI disscussed later in this book.

7.1.1 Basic definitions of stability

It is possible to introduce two simple definitions of stability:
- A system is *asymptotically stable* if and only if, in the neighborhood of an equilibrium point, all the eigenvalues associated to its linear part (or poles considered in the transfer function) are such that $Re(\lambda_i) < 0$ [9].

- A system is *stable* if and only if all the eigenvalues, in the neighborhood of an equilibrium point, all the eigenvalues associated to its linear part (or poles considered in the transfer function) are such that $Re(\lambda_i) \leqslant 0$. If there is an eigenvalue (or a pole) with a null real part, necessarily has geometric multiplicity equal to one [10].

It is also necessary to introduce a simple characterization of aeroelastic instability types:

- Static instability: *divergence*. It involves only steady aerodynamic and elastic forces. Mathematically it occurs when the eigenvalue crosses the imaginary axis with its only real part, which in turn becomes positive [11].
- Dynamical instability: *flutter*. It involves inertial, aerodynamical, and elastic forces. Mathematically it is described as a phenomenon when a couple of complex conjugate eigenvalues cross the imaginary axis acquiring a positive real part [12].

For a better understanding of the aeroelastic stability analysis, it is important to highlight the issues related to bifurcation problems.

7.2 Bifurcation problems

Let us start with an ordinary differential equation system depending on a k-dimensional parameter μ:

$$\dot{\mathbf{x}} = \mathcal{F}_\mu(\mathbf{x}) = \mathcal{F}(\mu, \mathbf{x}); \quad \mathbf{x} \in \mathbb{R}^n, \quad \boldsymbol{\mu} \in \mathbb{R}^k. \tag{7.1}$$

For the equilibrium state, we need

$$\mathcal{F}_\mu = 0. \tag{7.2}$$

Is possible to describe the equilibria varying the parameter μ with a smooth function using the implicit function theorem as long as the relation $x_{\text{eq}}(\mu)$ remains bi-univocal. When it is no more bi-univocal, different solutions coincide for different parameter sets. A lot of different mechanisms can characterize a bifurcation process, but in this work, we will only consider the dynamical one, called Hopf bifurcation where, for a critical value of the control parameter μ a stable equilibrium fixed point-like solution becomes unstable and a stable LCO arrises. It can be demonstrated that the statical or dynamical nature of this process depends on the spectrum behavior of the linearized system [13].

7.2.1 Hopf bifurcation

A Hopf bifurcation occurs when a fixed-point solution becomes linearly unstable beyond the critical value (μ_c) of a parameter μ. It could be of two types depending on the parameter value:

- It is called *subcritical* when $\mu < \mu_c$; before the bifurcation point the equilibrium stability is conditioned with an unstable LCO that defines the stability margin, and after the bifurcation point only an unstable equilibrium exists.
- It is called *supercritical* when $\mu > \mu_c$; after the bifurcation point there is an unstable equilibrium point and a stable LCO.

7.3 Aeroelastic problem formulation

Considering what has been said so far, we obtain the fundamental equation that describes the aeroelastic system, written in the Laplace transform domain:

$$(s^2 M + K)\tilde{q} = q_D E(s;\, U_\infty, M_\infty)\tilde{q} + \tilde{f}, \tag{7.3}$$

where, M, K, and E are the mass, stiffness, and GAF matrices, respectively; $\tilde{q}$ and $\tilde{f}$ are the Laplace transformed vectors of displacement and external force, respectively; q_D is the dynamic pressure, U_∞ the flow velocity, M_∞ is the Mach number, these parameters describe the flow. The stability problem can be solved using iterative methods, like the *p–k method* to the associated equation:

$$[s^2 M + K - q_D E]\tilde{w} = 0. \tag{7.4}$$

7.3.1 *p–k* method

The aeroelastic analysis carried out using SOL 145 implemented in MSC Nastran is based on the *p–k* method. Assume that the aerodynamics calculated for the pure harmonic motion is a good approximation for weakly damped ones. It is possible to affirm for a complex frequency $p = \alpha + jk \simeq jk$: neglecting the real part, i.e., the damping, whereas the transfer aerodynamic matrix is calculable. In this method further aerodynamic forces are considered as additional terms of stiffness and damping. Under this hypothesis Eq. (7.4) could be written as:

$$[s^2 M + sD + K - q_D[E_R(k;\, M_\infty) + jE_I(k;\, M_\infty)]]\tilde{w} = 0. \tag{7.5}$$

The GAF matrix is evaluated as a function defined in the Laplace's dimensionless sub-domain $k := \Im(s)l/U_\infty$ (i.e., Fourier's domain): the approximation becomes better closer to the imaginary axis.

Introducing the velocity vector $\tilde{v} := s\tilde{w}$ and defining

$$A_{\varrho_\infty, U_\infty, M_\infty; k} := $$

$$\begin{bmatrix} 0 & I \\ -M^{-1}\left[K - \frac{1}{2}\varrho_\infty U_\infty^2 E_R\left(k; M_\infty\right)\right] & -M^{-1}\left[D - \frac{1}{2}\varrho_\infty U_\infty b\frac{E_I\left(k; M_\infty\right)}{k}\right] \end{bmatrix} \tag{7.6}$$

and

$$\tilde{u} := \left\{ \begin{matrix} \tilde{w} \\ \tilde{v} \end{matrix} \right\}, \tag{7.7}$$

we obtain the following first-order problem:

$$\left(A_{\varrho_\infty, U_\infty, M_\infty; k} - sI\right)\tilde{u} = 0. \tag{7.8}$$

The implicit dependence on the complex eigenvalue s through k makes it necessary to resort to an iterative method.

7.3.1.1 Iterative method

1. Flight physics parameters U_∞, ϱ_∞, and M_∞ are fixed; in this way A matrix depends only on k.
2. *Divergence condition*
   Matrix A is evaluated for $k = 0$ and the divergence condition holds if the calculated eigenvalue is zero.
3. *Different from zero poles' calculation*
   The first structural mode frequency ω_1 is used in order to estimate the first approximation of k,

$$k_1^{(0)} := \frac{\omega_1 b}{U_\infty}.$$

It is possible to calculate the $2N$ problem eigenvalues $s_i^{(0)}$. Among these eigenvalues, $s_1^{(0)}$, the one with the imaginary part closest to ω_1, is the new estimated value of the reduced frequency,

$$k_1^{(1)} := \frac{s_1^{(0)} b}{U_\infty}.$$

Defining ε as the required accuracy of the solution, if the condition

$$\left\| k_1^{(1)} - k_1^{(0)} \right\| < \varepsilon$$

is satisfied, $s_1^{(0)}$ is the first aeroelastic pole, otherwise a new evaluation of the E matrix is required to calculate a new group of eigenvalues, iterating the process until the chosen convergence condition is satisfied.

Once the first pole is detected, the others could be calculated by the same iteration method using the associated structural frequency as the first attempt value. This method could also be used for varying the other parameters for the stability scenario.

7.4 Finite element method for flag-flutter

This section is devoted to numerical aspects. The numerical model analyzed for the aeroelastic stability of the system is implemented in SOL 145, a solver of software MSC Nastran, with which we consider a linear unsteady behavior for both the structure and aerodynamics.

7.4.1 Numerical model

The aeroelasticity phenomenon can be used for energy harvesting as well. For energy harvesting, piezoelectric materials can be a good option [14–19]. The carried out aeroelastic analysis is based on the use of both structural and aerodynamic finite elements. The aeroelastic analysis of the composite swept wings for an aircraft has been carried out in the literature by Polli et al. [20]. A finite element is defined by its geometry, while its physical displacement $u(x; t)$ is described by the functions of form $\phi^{(n)}(x)$, associated with the related nodes or grids:

$$u(x, t) \cong \sum_n^N \phi^{(n)}(x) q_n(t). \tag{7.9}$$

In particular, these are composed of panels, aligned in the direction of the steady (or harmonically unsteady) flow, on which the aerodynamic forces act. The mathematical model used, also known as *doublets lattice method* (or simply DLM), is based on the linearized theory of the aerodynamic potential. On each panel, there is a distribution of doublets equivalent to the pressure drop between the upper and lower surfaces of the panel itself.

Each doublet has an unknown intensity and is placed on the line 1/4 of the panel chord. Every element has a *collocation point* in the middle of the line placed at 3/4 of the chord in which the normal wash, $w = \nabla \phi \cdot \mathbf{n}$, is calculated. This is equal to the elastic oscillation of the structure itself.

Writing the equations for all the panels, we obtain a system of algebraic equations, which allow the calculation of doublets' intensity and, consequently, the pressure jump across the lifting surface. It is possible to calculate the aerodynamic forces acting on the surface itself. Since the structural grids do not usually coincide with the aerodynamic ones, it is necessary to make an interpolation between them.

This step is fundamental in order to generate the same deformation of the structure by two systems of distinct forces acting on the aerodynamic and structural grid points. This is achieved mostly by an equal amount of virtual work done by two force systems. Unsteady aerodynamic forces are described by the *GAF (generalized aerodynamic force)* matrix which relates the forces acting on the structure to its deformations. This GAF matrix is named the [E] matrix, which was introduced in Eq. (7.3). In the next chapters, the methodologies of piezoelectric energy harvesting by aeroelastic means are discussed in detail.

7.5 2D modeling for steady FSI system

In this section, structures and aerodynamics are studied separately and then combined together for the formulation of a fluid–structure interaction system [21].

7.5.1 2D structures

A 2D beam can be modeled as

$$\omega(x, y) = \iint C^{\omega p}(x, y; \xi, \eta) p(\xi, \eta) d\xi \, d\eta, \tag{7.10}$$

where ω is the vertical deflection at x and y, p is the pressure, and $C^{\omega p}$ is the deflection on x and y axis due to p. If

$$\omega(x, y) = h(y) + x\alpha(y) \tag{7.11}$$

and

$$C^{\omega p}(x, y; \xi, \eta) = C^{hF}(y, \eta) + x C^{\alpha F}(y, \eta) + \xi C^{hM}(y, \eta) + x\xi \, C^{\alpha M}(y, \eta) \tag{7.12}$$

132 Piezoelectric Aeroelastic Energy Harvesting

then ω and p are considered to be positive. Here C^{hF} is the deflection in y due to force F and $C^{\alpha F}$ is the twist due to force F. By substituting Eqs. (7.10), (7.11) and (7.12),

$$h(y) + x\alpha(y) = [\int C^{hF}(\int p(\xi, \eta)d\xi)\eta + \int C^{hM}(\int \xi p(\xi, \eta)d\xi)d\eta]$$
$$+ x[\int C^{\alpha F}(\int p(\xi, \eta)d\xi)d\eta + \int C^{\alpha M}(\int \xi p(\xi, \eta)d\xi)d\eta]. \tag{7.13}$$

If y and η are on same axis, resulting in $C^{hM} = C^{\alpha F}$, then

$$h(y) = \int C^{hF}(y, \eta)F(\eta)d\eta, \tag{7.14}$$

$$\alpha(y) = \int C\alpha M(y, \eta)M(\eta)d\eta \tag{7.15}$$

$$\text{where } F \equiv \int p \, d\xi \text{ and } M \equiv \int p\xi \, d\xi,$$

whereas Eq. (7.14) depicts that neither divergence nor control surface is affected by it. From Eq. (7.15), it can be observed that only M depends on α and it does not depend on h.

7.5.2 2D aerodynamic surfaces

By taking into account the aerodynamic forces affecting the deformation [21],

$$\frac{p(x, y)}{q} = \iint A^{p\omega x}(x, y; \xi, \eta)\frac{\partial \omega}{\partial \xi}(\xi, \eta)\frac{d\xi}{c_r}\frac{d\eta}{l}, \tag{7.16}$$

where $A^{p\omega x}$ is the pressure due to dimensionless aerodynamic forces, c_r and l are the reference chord and span, respectively.

If ω and p are considered to be positive then

$$\omega = h + x\alpha. \tag{7.17}$$

Therefore,

$$\frac{\partial \omega}{\partial x} = \alpha. \tag{7.18}$$

The lift for the beam can be defined as

$$L \equiv \int pdx = qc_r \int_0^1 A^{L\alpha}(y, \eta)\alpha(\eta)\frac{d\eta}{l} \tag{7.19}$$

and

$$A^{L\alpha} \equiv \iint A^{p\omega x}(x, y; \xi, \eta) \frac{d\xi}{c_R} \frac{dc}{c_r}. \tag{7.20}$$

7.5.3 FSI system

In this section, the matrix-lumped element approach is used for the solution by the integral approximation of matrix notations and summations [21]:

$$\{\omega\} = \Delta\xi \Delta\eta [C^{\omega p}]\{p\}. \tag{7.21}$$

Therefore, Eq. (7.16) becomes

$$\{p\} = q \frac{\Delta\xi}{c_r} \frac{\Delta\eta}{l} [A^{p\omega x}](\frac{\partial\omega}{\partial\xi}). \tag{7.22}$$

Now,

$$(\frac{\partial\omega}{\partial\xi}) \cong \frac{\omega_{i-1} - \omega_{i-1}}{2\Delta\xi}. \tag{7.23}$$

Hence, for the surface slope,

$$(\frac{\partial\omega}{\partial\xi}) = \frac{1}{2\Delta\xi}[W]\{\omega\} = \frac{1}{2\Delta\xi} \begin{bmatrix} [W] & [0] & [0] & [0] \\ & [W] & [0] & [0] \\ & & [W] & [0] \\ & & & [W] \end{bmatrix} \{\omega\}. \tag{7.24}$$

Eqs. (7.17)–(7.24) represent the matrix of numerical weighting. For ω, we have

$$[D]\{\omega\} \equiv [\begin{matrix} & 1 & \end{matrix}] - q\frac{(\Delta\xi)^2}{c_r} \frac{(\Delta\eta)^2}{l} \frac{l}{2\Delta\xi}[C^{\omega p}][A^{\omega px}][W]]\{\omega\} = \{0\}. \tag{7.25}$$

In case of divergence, $|D| = 0$, which helps in q_D determination.

7.6 Dynamic FSI

We realize from aerodynamics that the motion occurs by the "downwash" ω_α of the aerodynamic force, which is [21]

$$\omega_\alpha \equiv \frac{\partial z_a}{\partial t} + U_\infty \frac{\partial z_a}{\partial x}, \tag{7.26}$$

134 Piezoelectric Aeroelastic Energy Harvesting

where z_a represents the vertical displacement at time t. For an inviscid fluid, at the fluid–structure interface, e.g., at the surface of an airfoil, the boundary state involves an instantaneous location of the surface fluid portion which is normal to the surface to be equal with a normal surface speed. The fixed boundary condition in a coordinate system with regard to the fluid can be represented as

$$\omega_\alpha = \frac{\partial z_a}{\partial x}, \tag{7.27}$$

where ω_α represents the normal component of fluid velocity, and $\frac{\partial z_a}{\partial x}$ represents the surface's normal velocity. If the normal fluid total velocity becomes similar to the body normal velocity then

$$\omega_{\text{total}} = \omega_\alpha + \omega_G = \frac{\partial z_a}{\partial t} + U_\infty \frac{\partial z_a}{\partial x}, \tag{7.28}$$

where ω_G is the gust velocity in the vertical direction, which also affects the ω_α in the airfoil, and U_∞ represents the mean flow velocity. The airfoil's pressure loading can be represented as $p + p_G$ (p_G is the pressure due to ω_G), and the pressure p is due to

$$\omega_\alpha = -\omega_G(x, t) + \frac{\partial z_a}{\partial t} + U_\infty \frac{\partial z_a}{\partial x}. \tag{7.29}$$

The airfoil's loading pressure can be presented as

$$z_a = -h - \alpha x \quad \text{and} \quad \omega_\alpha = -\omega_G - \dot{h} - \dot{\alpha}x - U_\alpha \alpha. \tag{7.30}$$

By using aerodynamic theories, the lift and moment can be expressed as

$$L(t) \sim \int_{-\infty}^{\infty} I_{Lh}(t - \tau)[\dot{h}(\tau) + U_\infty \alpha(\tau)]d\tau + \int_{-\infty}^{\infty} I_{L\dot{\alpha}}(t - \tau)\dot{\alpha}(\tau)d\tau. \tag{7.31}$$

In dimensionless terms, the lift can be defined as

$$\frac{L}{qb} = \int_{-\infty}^{\infty} I_{Lh}(s - \sigma)[\frac{d\frac{h}{b}(\sigma)}{d\sigma} + \alpha(\sigma)]d\sigma + \int_{-\infty}^{\infty} I_{L\dot{\alpha}}(s - \sigma)[\frac{d\alpha(\sigma)}{d\sigma}]d\sigma. \tag{7.32}$$

In dimensionless terms, the moment can be defined as

$$\frac{M}{qb^2} = \int_{-\infty}^{\infty} I_{Mh}(s - \sigma)[\frac{d\frac{h}{b}(\sigma)}{d\sigma} + \alpha(\sigma)]d\sigma + \int_{-\infty}^{\infty} I_{M\dot{\alpha}}(s - \sigma)[\frac{d\alpha(\sigma)}{d\sigma}]d\sigma. \tag{7.33}$$

If a typical section is considered, $s \equiv \frac{tU_\infty}{b}$, $\sigma \equiv \frac{\tau U_\infty}{b}$, and $I_{L\dot{h}}$ represents the impulse function due to aerodynamics which is dependent on the Mach number. The Fourier transform of Eq. (7.33) can be presented as [21]

$$\frac{\bar{L}(k)}{qb} \equiv \int_{-\infty}^{\infty} \frac{L(s)}{qb} e^{-iks} ds = \int_{-\infty}^{\infty} \int_{-\infty}^{\infty} I_{L\dot{h}}(s-\sigma)[\frac{d\frac{h}{b}}{d\sigma} + \alpha] e^{-iks} d\sigma \, ds + \cdots .$$

$$(7.34)$$

The reduced frequency k is equivalent to $\frac{\omega b}{U_\infty}$. Let

$$\gamma \equiv s - \sigma \quad \text{and} \quad d\gamma = ds. \tag{7.35}$$

Then Eq. (7.34) becomes

$$\frac{\bar{L}(k)}{qb} = \int_{-\infty}^{\infty} \int_{-\infty}^{\infty} I_{L\dot{h}}(\gamma)[\frac{d\frac{h}{b}}{d\sigma} + \alpha] e^{-ik\gamma} d\sigma \, d\gamma + \cdots = H_{L\dot{h}}(k)[ik\frac{\bar{h}}{b} + \bar{\alpha}] + \cdots ,$$

$$(7.36)$$

$$H_{L\dot{h}}(k) \equiv \int_{-\infty}^{\infty} I_{L\dot{h}}(\gamma) e^{-ik\gamma} d\gamma,$$

$$\frac{\bar{h}}{b} \equiv \int_{-\infty}^{\infty} \frac{h(\sigma)}{b} e^{-ik\sigma} d\sigma,$$

$$\bar{\alpha} \equiv \int_{-\infty}^{\infty} \alpha(\sigma) e^{-ik\sigma} d\sigma.$$

Then Eqs. (7.32) and (7.36) become

$$\frac{\bar{L}}{qb} = H_{L\dot{h}}[ik\frac{\bar{h}}{b} + \bar{\alpha}] + H_{L\dot{\alpha}} ik\bar{\alpha}, \tag{7.37}$$

$$\frac{\bar{M}_y}{qb^2} = H_{M\dot{h}}[ik\frac{\bar{h}}{b} + \bar{\alpha}] + H_{M\dot{\alpha}} ik\bar{\alpha}, \tag{7.38}$$

while considering the classical aerodynamic approach, the coordinate system origin is generally taken in the middle chord. Therefore,

$$z_\alpha = -h - \alpha(x - x_{x_{e.a.}}), \tag{7.39}$$

$$\omega_\alpha = -\dot{h} - \dot{\alpha}(x - x_{x_{e.a.}}) - U_\infty \alpha$$
$$= (-\dot{h} - U_\infty \alpha) - \dot{\alpha}(x - x_{x_{e.a.}})$$
$$= \dot{h} - U_\alpha \alpha + \dot{\alpha} x_{e.a.} - \dot{\alpha} x.$$

Table 7.1 Aerodynamic models' summary.

Mach number	2D Geometry	3D Geometry
$M \ll 1$	Yes	For transfer function determination
$M \approx 1$	Limited	Linear, inviscid transfer functions and nonlinear and/or viscous effects
$M \gg 1$	Yes and simple	Yes and simple

In Table 7.1, we will outline aerodynamic hypotheses in terms of the Mach number and geometry, typically as seen in industrial practices. All this assumes that flow models are inviscid and minor disturbances are potential.

7.7 Aerodynamic theories

As per Table 7.1, if $M \gg 1$ then the pressure of loading acting on an airfoil is [22]:

$$p = \rho \frac{U_\infty^2}{M} [\frac{\frac{\partial z_a}{\partial t} + U_\infty \frac{\partial z_a}{\partial x}}{U_\infty}]$$
$$= \rho a_\infty [\frac{\partial z_a}{\partial t} + U_\infty \frac{\partial z_a}{\partial x}].$$

(7.40)

The pressure at position x and y at time t only depends on the movement at the same position and time and does not rely on the movement at other positions, i.e., the local influence or the movement at previous times (zero memory effect). Some of these are called the aerodynamic "piston principle" since the friction is on a piston in a speed tunnel:

$$\omega_a = \frac{\partial z_a}{\partial t} + U_\infty \frac{\partial z_a}{\partial x}.$$

(7.41)

This velocity–pressure relationship has been extensively used in aeroelasticity in recent years and is also popular for its 1D wave acoustic theory. The aerodynamic "piston principle" allows for impulses and transferrals. Particularly established, but not as simple as those for $M \gg 1$, are "aerodynamic moment function" and "aerodynamic transfer functions" for a 2D incompressible stream. The lift because of transient motion is usually written as

$$\frac{L}{qb} = 2\pi\left[\frac{d^2\frac{h}{b}}{ds^2} + \frac{d\alpha}{ds} - a\frac{d^2\alpha}{ds^2}\right]$$
$$+ 4\pi\{\phi(0)[\frac{d\frac{h}{b}}{ds} + \alpha + (\frac{1}{2} - a)\frac{d\alpha}{ds}] \quad (7.42)$$
$$\cdot \int_0^s (\frac{d\frac{h}{b}}{d\sigma} + \alpha + (\frac{1}{2} - a)\frac{d\alpha}{ds})\dot{\phi}(s - \sigma)d\sigma\}.$$

From Eqs. (7.32) and (7.42), we get

$$I_{Lh} = 2\pi D + 4\pi\dot{\phi} + 4\pi\phi(0)\delta, \quad (7.43)$$
$$I_{L\dot{\alpha}} = 4\pi(\frac{1}{2} - a)\dot{\phi} + 4\pi(\frac{1}{2} - a)\phi(0)\delta - 2\pi aD, \quad (7.44)$$

where δ represents the delta function while D represents the derivative of δ known as doublet function. In fact, Eq. (7.42) is used instead of Eq. (7.32) since δ and D are not appropriate to be numerically integrated, etc. However, the formal equivalent of Eqs. (7.42) and (7.32) are Eqs. (7.43) and (7.44). Notice that Eq. (7.42) is more suitable in physical terms as well. The downwash at the $\frac{3}{4}$ of the chord can be expressed as $-[\frac{d\frac{h}{b}}{ds} + \alpha + (\frac{1}{2} - a)\frac{d\alpha}{ds}]$. Therefore, $\frac{3}{4}$ of the chord is very important for the 2D flow which is incompressible.

7.7.1 Wagner function

Functions I_{Lh} and $I_{L\dot{\alpha}}$ (the impulses due to aerodynamics) can be fully expressed in the single function ϕ, known as Wagner function [23]. The Wagner function is represented in Fig. 7.2. The approximation can be expressed as

$$\phi(s) = 1 - 0.615e^{-0.0455s} - 0.335e^{0.3s}. \quad (7.45)$$

Figure 7.2 Wagner function.

The flow smooths the flow of D and δ functions of Eqs. (7.43) and (7.44) with Mach numbers higher than zero, and no simple form like Eq. (7.42) occurs. Therefore, Eq. (7.42) is especially useful for an incompressible flow. The gust velocity in the vertical direction, w_G, is dependent on an analogous impulse function. The lift and moment can be described as

$$\frac{L_G}{qb} = \oint_{-\infty}^{\infty} I_{LG}(s-\sigma)\frac{w_G(\sigma)}{U}d\sigma, \qquad (7.46)$$

$$\frac{M_{yG}}{qb^2} = \int_{-\infty}^{\infty} I_{MG}(s-\sigma)\frac{w_G(\sigma)}{U}d\sigma. \qquad (7.47)$$

7.7.2 Kussner function

In the case of an incompressible flow,

$$I_{LG} = 4\pi\dot{\psi}, \qquad (7.48)$$

$$I_{MG} = I_{LG}(\frac{1}{2}+a), \qquad (7.49)$$

where ψ represents the Kussner function [24]. The Kussner function is represented in Fig. 7.3 and expressed in Eq. (7.50) as

$$\psi(s) = 1 - 0.5e^{-0.13s} - 0.5e^{-s}. \qquad (7.50)$$

For transient aerodynamic loading of airfoils, the Wagner and Kussner functions are commonly used. They are also employed with analytical corrections to the Mach number effects even with a compressible, subsonic flow. For two-dimensional, supersonic flow, too, relatively simple, exact formulae exist. However, aerodynamic impulse functions should be calculated with very elaborative computational means in the case of subsonic

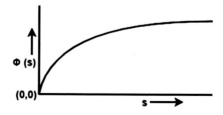

Figure 7.3 Kussner function.

and/or 3D flow. So, Eqs. (7.46) and (7.47) can be expressed in the frequency domain as

$$\frac{\bar{L}_G}{qb} = H_{LG}(\omega)\frac{\bar{\omega}_G}{U}, \tag{7.51}$$

$$\frac{\bar{M}_{yG}}{qb^2} = H_{MG}(\omega)\frac{\bar{\omega}_G}{U}. \tag{7.52}$$

Eqs. (7.51) and (7.52) are useful when using power spectral methods to approach the gust problem in random procedures.

7.8 General approximation

The spatial or temporal dependency of the aerodynamic forces is often simplified in assumptions. Three commonly used methods are discussed in this section. The following are the assumption theories.

7.8.1 Strip theory

The established results for a 2D flow (infinite airfoil span) are used in this approximation to quantify the aerodynamic forces on a finite span lifting surface. It is essential to take any span-wise station into account as if it was a portion of a long wing with standardized span-specific characteristics. The elevation (or, more precisely, the chord-wise pressure distributor) at any span-by-span station is therefore supposed to rely only on the aerodynamic principle at this station and to be free from the downwashing at any other span-by-station [25].

7.8.2 Quasi-steady

The above-described strip theory approach is unambiguous and is widely agreed upon in its significance. This, unfortunately, is not valid for the approximation that is almost constant. Its qualitative importance is widely agreed, i.e., the temporary memory effect of the aerodynamic model is ignored and the aerodynamic forces are only at that time contingent on the movement of the airfoil and independent of the movements in previous periods. That is, as far as aerodynamic forces are concerned, the history of the transfer is ignored. The piston principle, for example, is a quasi-steady approximation [26].

For example, the aerodynamic forces for a 2D incompressible flow can be taken into consideration by building an almost constant approximation.

140   Piezoelectric Aeroelastic Energy Harvesting

Often, one such solution is to approximate the Wagner function by

$$\phi = 1 \quad \text{then} \quad \phi(0) = 1 \quad \text{and} \quad \dot{\phi} = 0.$$

It is certainly a quasi-steady model as the integral convolution in Eq. (7.42) can now be measured with respect to the actual motion of the airfoil $s \equiv \frac{tU_\infty}{b}$, and the aerodynamic forces are now independent of airfoil motion's history.

An alternative sometimes used quasi-steady method is first to obtain the aerodynamic forces to sustain a steady motion, for example, only certain terms that contain α in (7.42), and then to set the corresponding unstable angle of attack,

$$\alpha + \frac{dh}{dt}\frac{1}{U_\infty},$$

substituting α in the steady aerodynamic theory. This second estimate of quasi-steady clearly varies from the previous. The uncertainty could be avoided if a general consent exists, namely that an extension of the reduced frequency for sinusoidal airfoil motion is intended by a quasi-steady approximation. Nevertheless, consensus can also be found on the number of requirements to be maintained through expansion.

7.8.3 Slender body/wing

Another spatial approximation is possible if the lifting surface is of a low aspect ratio or if a slender body is involved. In these situations, in comparison with span-wise changes, the chord-wise spatial rate of change (derivatives) should be disregarded and the chord-wise coordinate then becomes an important parameter and not an independent coordinate [27]. For slim bodies and lower aspect ratio wings, the asymptotic check of numerical methods is helpful. However, only a modest set of functional lifting surfaces is beneficial for quantitative forecasts. For an outward flow around a slender body, a particularly fascinating finding is available if the body has rigid cross-sections which deform only in the direction of the body:

$$z_\alpha(x, y, t) = z_a(x, t). \tag{7.53}$$

The lifting force per unit provided by chord distance is

$$L = -\rho_\infty \frac{dS}{dx} U[U\frac{\partial z_a}{\partial x} + \frac{\partial z_a}{\partial t}] - \rho_\infty S[U^2\frac{\partial^2 z_a}{\partial x^2} + 2U\frac{\partial^2 z_a}{\partial x \partial t} + \frac{\partial^2 z_a}{\partial t^2}]. \tag{7.54}$$

In a compact form, for a cylinder having circular cross-section constant ($s = \pi R^2$ and $\frac{dS}{dx} = 0$), Eq. (7.54) can be written as

$$L = -\rho_\infty S[U^2 \frac{\partial^2 z_a}{\partial x^2} + 2U \frac{\partial^2 z_a}{\partial x \partial t} + \frac{\partial^2 z_a}{\partial t^2}]. \qquad (7.55)$$

Eq. (7.54) represents the internal flow by the lift force.

References

[1] Raymond L. Bisplinghoff, Holt Ashley, Robert L. Halfman, Aeroelasticity, Courier Corporation, 2013.

[2] Jan Robert Wright, Jonathan Edward Cooper, Introduction to Aircraft Aeroelasticity and Loads, vol. 20, John Wiley & Sons, 2008.

[3] Hassan Elahi, Zubair Butt, Marco Eugnei, Paolo Gaudenzi, Asif Israr, Effects of variable resistance on smart structures of cubic reconnaissance satellites in various thermal and frequency shocking conditions, Journal of Mechanical Science and Technology 31 (9) (2017) 4151–4157.

[4] Hassan Elahi, Marco Eugeni, Paolo Gaudenzi, Faisal Qayyum, Raees Fida Swati, Hayat Muhammad Khan, Response of piezoelectric materials on thermomechanical shocking and electrical shocking for aerospace applications, Microsystem Technologies 24 (9) (2018) 3791–3798.

[5] Hassan Elahi, Marco Eugeni, Paolo Gaudenzi, Madiha Gul, Raees Fida Swati, Piezoelectric thermo electromechanical energy harvester for reconnaissance satellite structure, Microsystem Technologies 25 (2) (2019) 665–672.

[6] Hassan Elahi, Khushboo Munir, Marco Eugeni, Muneeb Abrar, Asif Khan, Adeel Arshad, Paolo Gaudenzi, A review on applications of piezoelectric materials in aerospace industry, Integrated Ferroelectrics 211 (1) (2020) 25–44.

[7] Raymond L. Bisplinghoff, Holt Ashley, Principles of Aeroelasticity, Courier Corporation, 2013.

[8] Wei Shyy, Hikaru Aono, Satish Kumar Chimakurthi, Pat Trizila, C-K. Kang, Carlos E.S. Cesnik, Hao Liu, Recent progress in flapping wing aerodynamics and aeroelasticity, Progress in Aerospace Sciences 46 (7) (2010) 284–327.

[9] Earl H. Dowell, Panel flutter – a review of the aeroelastic stability of plates and shells, AIAA Journal 8 (3) (1970) 385–399.

[10] M.H. Hansen, Aeroelastic stability analysis of wind turbines using an eigenvalue approach, Wind Energy: An International Journal for Progress and Applications in Wind Power Conversion Technology 7 (2) (2004) 133–143.

[11] Hamed Haddad Khodaparast, John E. Mottershead, Kenneth J. Badcock, Propagation of structural uncertainty to linear aeroelastic stability, Computers & Structures 88 (3–4) (2010) 223–236.

[12] Carl Scarth, Jonathan E. Cooper, Paul M. Weaver, Gustavo H.C. Silva, Uncertainty quantification of aeroelastic stability of composite plate wings using lamination parameters, Composite Structures 116 (2014) 84–93.

[13] Ken J. Badcock, Mark A. Woodgate, Bryan E. Richards, Direct aeroelastic bifurcation analysis of a symmetric wing based on the Euler equations, Journal of Aircraft 42 (3) (2005) 731–737.

[14] Mohsin Ali Marwat, Weigang Ma, Pengyuan Fan, Hassan Elahi, Chanatip Samart, Bo Nan, Hua Tan, David Salamon, Baohua Ye, Haibo Zhang, Ultrahigh energy density and thermal stability in sandwich-structured nanocomposites with dopamine@Ag@BaTiO$_3$, Energy Storage Materials 31 (2020) 492–504.

[15] Ahsan Ali, Riffat Asim Pasha, Hassan Elahi, Muhammad Abdullah Sheeraz, Saima Bibi, Zain Ul Hassan, Marco Eugeni, Paolo Gaudenzi, Investigation of deformation in bimorph piezoelectric actuator: analytical, numerical and experimental approach, Integrated Ferroelectrics 201 (1) (2019) 94–109.

[16] Muhammad Usman Khan, Zubair Butt, Hassan Elahi, Waqas Asghar, Zulkarnain Abbas, Muhammad Shoaib, M. Anser Bashir, Deflection of coupled elasticity–electrostatic bimorph PVDF material: theoretical, FEM and experimental verification, Microsystem Technologies 25 (8) (2019) 3235–3242.

[17] Vittorio Memmolo, Hassan Elahi, Marco Eugeni, Ernesto Monaco, Fabrizio Ricci, Michele Pasquali, Paolo Gaudenzi, Experimental and numerical investigation of PZT response in composite structures with variable degradation levels, Journal of Materials Engineering and Performance 28 (6) (2019) 3239–3246.

[18] H. Elahi, A. Israr, R.F. Swati, H.M. Khan, A. Tamoor, Stability of piezoelectric material for suspension applications, in: 2017 Fifth International Conference on Aerospace Science & Engineering (ICASE), IEEE, 2017, pp. 1–5.

[19] Hassan Elahi, Marco Eugeni, Paolo Gaudenzi, Electromechanical degradation of piezoelectric patches, in: Analysis and Modelling of Advanced Structures and Smart Systems, Springer, 2018, pp. 35–44.

[20] Gian Mario Polli, Liviu Librescu, Franco Mastroddi, Aeroelastic response of composite aircraft swept wings impacted by a laser beam, AIAA Journal 44 (2) (2006) 382–391.

[21] Earl H. Dowell, Howard C. Curtiss, Robert H. Scanlan, Fernando Sisto, A Modern Course in Aeroelasticity, vol. 3, Springer, 1989.

[22] Howard C. Curtiss Jr, Robert H. Scanlan, Fernando Sisto, A Modern Course in Aeroelasticity, vol. 11, Springer Science & Business Media, 2013.

[23] Sh Shams, MH Sadr Lahidjani, H. Haddadpour, Nonlinear aeroelastic response of slender wings based on Wagner function, Thin-Walled Structures 46 (11) (2008) 1192–1203.

[24] Carlo L. Bottasso, Alessandro Croce, Federico Gualdoni, Pierluigi Montinari, Load mitigation for wind turbines by a passive aeroelastic device, Journal of Wind Engineering and Industrial Aerodynamics 148 (2016) 57–69.

[25] Dae-Kwan Kim, Jun-Seong Lee, Jin-Young Lee, Jae-Hung Han, An aeroelastic analysis of a flexible flapping wing using modified strip theory, in: Active and Passive Smart Structures and Integrated Systems 2008, vol. 6928, International Society for Optics and Photonics, 2008, 69281O.

[26] William P. Rodden, J. Richard Love, Equations of motion of a quasisteady flight vehicle utilizing restrained static aeroelastic characteristics, Journal of Aircraft 22 (9) (1985) 802–809.

[27] Rafael Palacios, Carlos Cesnik, Static nonlinear aeroelasticity of flexible slender wings in compressible flow, in: 46th AIAA/ASME/ASCE/AHS/ASC Structures, Structural Dynamics and Materials Conference, 2005, p. 1945.

CHAPTER 8

Flutter-based aeroelastic energy harvesting

Contents

8.1.	Introduction	143
8.2.	Flutter analysis: classical approach	144
8.3.	Flutter solutions	146
8.4.	Aeroelastic energy harvesters based on flutters	147
References		153

8.1 Introduction

The flutter mechanism is a dynamic instability typical of aeroelastic systems which can be helpful for energy harvesting purpose. Indeed, it individuates the borders of stability, e.g., with respect the flow velocity, of the unperturbed solution of aeroelastic system: after them our system experience a persistent and self-induced motion from which thanks to suitable transducers energy can be extracted. Here in the following some historical remarks about flutter and its study are provided.

Aeroelasticity and related phenomena are important aspects in aircraft design Garrick and Reed [1] also found an outstanding historical analysis of the evolution of aeroelasticity up to this time. Collar [2] gave a groundbreaking paper that marked the beginning of the field in 1946. A noteworthy analysis was also carried out by Collar, underlining British aeroelasticity efforts until that time [3]. Most complex data was taken at that moment with the oscilloscopes and the speed needed to be carefully calibrated to capture the frequency of interest.

In addition, the results could be real-time evaluated statistically with ongoing experiments. During the mid-1950s, Bisplinghoff discussed the influence of aerodynamic heating on the aircraft structure at high flight speed [4]. Furthermore, he discovered that, due to thermal stress and reduced material properties, the structure could be handled without rigidity. Additional observations into the flutter process emerged at the end of the 1950s. The basic bending–torsional flutter criteria of the standard section confirm that this flutter will occur even though the coalescence of bending and torsion frequencies does not cause damping forces [5,6].

Piezoelectric Aeroelastic Energy Harvesting
https://doi.org/10.1016/B978-0-12-823968-1.00019-2

Copyright © 2022 Elsevier Inc.
All rights reserved.

143

By the mid-1970s, aeroelastic control system problems had been investigated in a refined state. Aircraft flutter research has also been reformulated from the frequency domain to the more versatile time domain to support control systems in the time domain [7]. Aeroelasticity has fairly rapidly developed during the sixth decade of the flight. The analytical approach usually involved a flutter analysis of two, three, or four degrees of freedom (DoF), focusing on separate aspects of the wing and tail. Methods have usually been grouped into effective matrix formulations, and the simulations were conducted by a community of mechanical desk computers. During this time, the characteristics of the control system were devised and conveniently integrated into the aircraft flutter programs [8,9]. A variety of wind tunnel experiments on small and medium-sized models have demonstrated the principle of wing cross-section warping in order to reduce the gust response and expand the flutter boundary [10]. In many aeroelastic behaviors, non-linearities are normal, in particular in the case of steady oscillations. In the frequency domain, some initial cases of non-linear effects of flutter have been investigated using functions to model free play or Coulomb friction effects in control systems [11,12]. Subsequent studies included the implications of aerodynamic stall effects on flutter [6], large static wing deflections [13,14], and in the regime of transonic flow [15]. It is possible to explore the non-linear phenomenon in the time domain, as historically applied, by increasing processing power. The flutter mechanism is widely used currently for energy harvesting purposes by using piezoelectric materials [16–21].

8.2 Flutter analysis: classical approach

The aircraft industry carried out most of the lift-surface flutter analyses based on what is generally referred to as the "classical flutter analysis" on the basis of the determinant flutter till the end of the 1970s. The objective of such an examination was to determine the flight conditions which concern the flutter borders. The flutter border was previously noted as being a state on which a mode of movement is dependent on harmonic time. Since this is called a stability limit, it is presumed that motion modes are converging (i.e., stable) in less critical flight conditions (i.e., lower airspeed). In addition, apart from critical modes, all other modes are converging at the boundary of the flutter. The system of research is not dependent on the solution of generalized motion equations, rather, it is thought that the solution requires basic harmonic motion. The motion equations for the flight condition(s) which create such a solution are then determined.

In the p-system, we calculate the values for the set flight conditions, of which the modal damping is the precise element. Classical flutter cannot be clearly analyzed when subjected to modal damping for subjective flight conditions. Therefore, no other definite flutter stability indicator than the stability boundary position can be provided. While this is the primary drawback of a system of this kind, the primary strength is that only unsteady air loads are required for the basic harmonic motion of the surface, which is more readily extracted for a given degree of precision than for arbitrary motion.

In order to demonstrate the classical flutter analysis, it is important to consider the appropriate representation of non-steady air loads for basic harmonic movement of the lifting surface. Because these oscillatory movements are comparatively small in amplitude, this load can be calculated using the linear aerodynamic theory. These aerodynamic ideas are generally based on the linear flow theory of thin airfoils, which implies that the wing structure thickness and motion cause a slight disruption in the field of flow, and the velocity of flow disturbance is small compared to the free flow velocity. For demonstration purposes, it is necessary to reassess the standard portion of a 2D lifting surface which undergoes simultaneous rotational and translational movements. The motion is simple harmonic; therefore, g and ϕ are represented as simple harmonics:

$$g(t) = \bar{g}\exp(i\omega t), \tag{8.1}$$

$$\phi(t) = \bar{\phi}\exp(i\omega t), \tag{8.2}$$

where ω denotes the motion frequency. Even though ϕ and g are the same frequency motion, they are not usually in phase. This can be mathematically taken into consideration by taking the amplitude $\bar{\phi}$ as being real and $\bar{g}$ as a complex value. Since the linear aerodynamic principle is to be used, the corresponding lift, L, and the pitching moment around P, denoted by M, related by

$$M = M_{\frac{1}{4}} + b\left(\frac{1}{2} + a\right)L, \tag{8.3}$$

are simple harmonic too with frequency ω such that

$$M(t) = \bar{M}\exp(i\omega t), \tag{8.4}$$

$$L(t) = \bar{L}\exp(i\omega t). \tag{8.5}$$

146 Piezoelectric Aeroelastic Energy Harvesting

Computation of the air load amplitude can be done using complex linear functions:

$$\bar{L} = -\pi \rho_\infty b^3 \omega^2 [l_g(K, M_\infty)\frac{\bar{g}}{b} + l_\phi(K, M_\infty)\bar{\phi}], \qquad (8.6)$$

$$\bar{M} = \pi \rho_\infty b^4 \omega^2 [m_g(K, M_\infty)\frac{\bar{g}}{b} + m_\phi(K, M_\infty)\bar{\phi}]. \qquad (8.7)$$

Here ρ_∞ denotes the free stream air density. The four complex functions contained in the square brackets feature the aerodynamic dimensionless coefficients for a moment and lift arising from plunge and pitch. In general, these coefficients are the two parameters functions M_∞ (free stream Mach) and K (reduced frequency):

$$M_\infty = \frac{U}{c_\infty}, \qquad (8.8)$$

$$K = \frac{b\omega}{U}. \qquad (8.9)$$

As with steady air loads, the effects of compressibility are expressed in the dependency of the coefficients on M_∞. The reduced frequency K is special to the unsteady flow. Also, note that every coefficient does enter as a complex number for any given values of K and M_∞. The place where the flutter happens corresponds to the unique K and M_∞ values and requires iterative calculation.

8.3 Flutter solutions

The primary argument for defining simple harmonic time dependence is, of course, its correspondence with the boundary of stability. The primary argument for defining the time dependence of a simple harmonic is, for sure, corresponding with the stability boundary. This form of solution can be credited to Theodorsen (1934), who provided the first detailed flutter study with the production of unsteady air loads on a two-dimensional wing with an incompressible potential flow. While unsteady aerodynamic analyses for basic harmonic motion are not easy to devise and implement, they are much more workable than those for oscillatory motion analysis varying amplitudes. Development of several unsteady aerodynamic formulations has been done after Theodorsen's work for the simple harmonic motion for surface lifting. These methods have proved to be sufficient for compressible flows in both supersonic and subsonic regimes. They were also designed for

3D surfaces and for surface-to-surface interactions. These are very critical issues for aerospace engineering [22–25].

There are two other significant factors for the working engineer. The first factor is the development of the understanding of a stability margin under flight conditions near the flutter border. The second, and perhaps the most important, is to develop an understanding of the physical system that induces instability. With these two pieces of a puzzle, the engineer may suggest design variations that may minimize or even remove instability. When an adequate aerodynamic theory is available, the p-approach will answer these concerns. In this section, we look at alternative methods, in which engineers have solved these issues, where unstable aerodynamic theories that presume basic harmonic motion have to be used.

8.4 Aeroelastic energy harvesters based on flutters

Elahi et al. [26] performed a numerical and experimental campaign to analyze the aeroelastic energy harvesting based on LCOs. The overall mechanism is represented in Fig. 8.1. In this campaign, the fiberglass flag attached with a piezoelectric patch and Al patch were subjected to axial flow in a subsonic wind tunnel. When the airflow was increased up to a certain point, i.e., flutter velocity, the harvester started flapping and diverged from LCOs. The bifurcation diagrams for the Al and PZT patched harvester are represented in Figs. 8.2 and 8.3, respectively.

When the harvester (of different lengths) was subjected to axial airflow, the flutter velocity and flutter frequency varied as shown in Fig. 8.4. It can be observed that with the increase in the length of the harvester the flutter velocity and frequency reduces. It means that with the harvester of greater length it will be easier to set it to the limit cycle of oscillations [27].

For the last few decades, the concept of energy harvesting from aeroelastic vibrations has become popular [26–32]. Certain research focused on aeroelastic energy harvesters based on an airfoil that use a piezoelectric transducer. The possibility of taking in a piezoelectric power harvester from a flow-excited morphing airfoil with MFCs as a starting point for energy harvesting from morphing aircraft MFCs has also been presented at Erturk et al. [33]. They used 133 mm span and 127 mm chord of the airfoil. Airfoil was made of a bimorph flat plate having four piezoceramic patches M 8557 P1 MFC, two of them were on the top and two at the bottom of the surface, whereas the used material for the substrate was stainless steel. For the

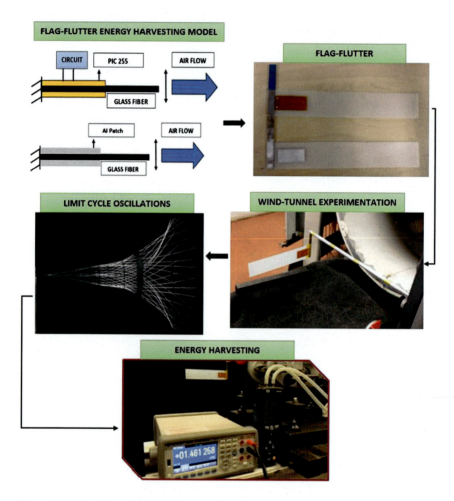

Figure 8.1 Overall mechanism of piezoelectric energy harvesters [26].

generation of power, the authors used only two PZ patches on the top of the surface. The resultant max RMS power of this experiment was 7 µW at 20° of attack angle, with velocity equal to 15 m/s and 98 kΩ electrical resistance.

The authors of [29] assumed that the wing behavior can be accurately interpreted by using just the pitch, plunge, and flap mode. They hypothesized that the wing's structural dynamics do not get influenced by the piezoelectric patch, i.e., no modification in modes and stiffness properties. The charge on the surface of a piezoelectric material is dependent on plunge motion and is influenced by the bending experience.

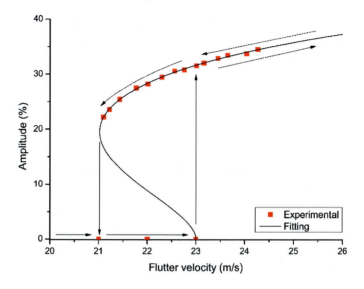

Figure 8.2 Bifurcation diagram of the Al patched harvester [26].

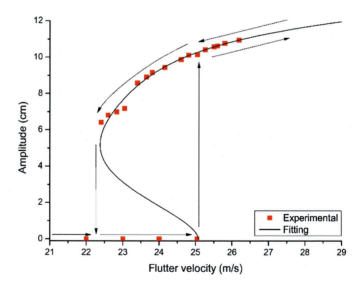

Figure 8.3 Bifurcation diagram of the aeroelastic piezoelectric energy harvester [26].

Theodorsen in 1935 described the aeroelastic behavior, in which he studied the structural system via the well-known typical section airfoil model, as given in the equations below:

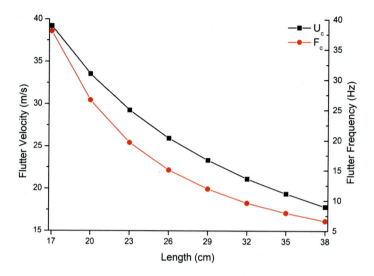

Figure 8.4 Response to flutter velocity and frequency by the harvester of various length.

$$\ddot{\xi} + x_\alpha \ddot{\alpha} + x_\beta \ddot{\beta} + \frac{\Omega_1^2}{U^2}\xi = p, \quad (8.10)$$

$$\frac{x_\alpha}{r_\alpha^2}\ddot{\xi} + \ddot{\alpha} + [r_\beta^2 + (c_h - a_h)x_\beta]\frac{1}{r_\alpha^2}\ddot{\beta} + \frac{1}{U^2}M_\alpha(\alpha) = r, \quad (8.11)$$

$$\frac{x_\beta}{r_\beta^2}\ddot{\xi} + \ddot{\beta} + [1 + (c_h - a_h)\frac{x_\beta}{r_\beta^2}]\ddot{\alpha} + \frac{\Omega_2^2}{U^2}M_\beta(\beta) = s, \quad (8.12)$$

where

$$M_\alpha(\alpha) = c_{1\alpha}\alpha, \quad (8.13)$$

$$M_\beta(\beta) = \frac{-2c_{1\beta}\delta}{\pi}\arctan(\frac{\pi\beta}{2\delta}) + c_{1\beta}\beta. \quad (8.14)$$

They considered the approximation of arctangent for computational simplification, which is rich enough for covering all responses except from the edges. A control surface trailing-edge is modeled by considering a spring that is located on the flap hinge. In the equation written above, the differentiation with respect to time τ is represented by overdots such that $\tau = Vt/b$, the dimensionless plunge displacement is denoted by $\xi = h/b$, $r_\alpha = \sqrt{J_\alpha/mb^2}$ is gyration's reduced radius of the elastic axis, $r_\beta = \sqrt{J_\beta/mb^2}$ is gyration's reduced radius of flap hinge, and mass ratio is given by $\mu = \sqrt{\pi\rho b^2/m}$. The symbols Ω_2 and Ω_1 are given by $\Omega_2 = \omega_\beta/\omega_\alpha$

and $\Omega_1 = \omega_\xi / \omega_\alpha$, where ω_α ω_ξ, and ω_β are the pitching, uncoupled plunging, and flapping natural frequencies. The wing section's torsional spring moment's overall contribution is represented as $M_\alpha(\alpha)$ where $M_\beta(\beta)$ is the overall contribution including both non-linear and linear parts of torsional spring moment of the wing section. The reduced flow velocity is $U = V/b\omega_\alpha$, where V is speed. We also have the following equations:

$$p = -\frac{1}{\mu}\,[\dot\alpha + \ddot\xi - a_h\ddot\alpha - (T_4/\pi)\dot\beta - (T_1/\pi)\ddot\beta - 2u(w_{3/4})\,],$$

$$\begin{aligned}
r = &-\frac{1}{\mu r_\alpha^2}[\bar{a}_h\dot\alpha + (1/8 + a_h^2)\ddot\alpha - a_h\ddot\xi + (T_4/\pi + T_{10}/\pi)\beta \\
&+ (T_1/\pi - T_8/\pi - (c_h - a_h)T_4/\pi + 1/2(T_{11}/\pi))\dot\beta \\
&- (T_7/\pi + (c_h - a_h)T_1/\pi)\ddot\beta - 2(a_h + 1/2)u(w_{3/4})\,],
\end{aligned}$$

$$\begin{aligned}
s = &-\frac{1}{\mu r_\beta^2}[(-2T_9/\pi - T_1/\pi - \bar{a}_h T_4/\pi)\dot\alpha + 2(T_{13}/\pi)\ddot\alpha \\
&+ (T_5/\pi^2 - T_4 T_{10}/\pi^2)\beta - (T_4/\pi)(T_{11}/\pi)\dot\beta \\
&- (T_3/\pi^2)\ddot\beta - (T_1/\pi)\ddot\xi + (T_{12}/\pi)u(w_{3/4})],
\end{aligned}$$

$$\text{(8.15)}$$

where r represents the pitch and p represents the lift of the aerodynamic moment for wing, while s is the flap's pitching moment [34]. The circulatory part of pitching moments and lift represented with $u(w_{3/4})$ is taken into account as $\int_0^\tau \phi(\tau - \sigma)\dot{w}_{3/4}(\sigma)d\sigma$, where

$$w_{3/4}(\tau) = \dot\xi(\tau) + \tilde{a}_h\dot\alpha(\tau) + \alpha(\tau) + (T_{10}/\pi)\dot\beta + (T_{11}/\pi)\ddot\beta \qquad \text{(8.16)}$$

$$\tilde{a}_h = 1/2(1 - a_h), \qquad \text{(8.17)}$$

where the chord's third-quarter displacement is represented by w and Wagner function is given by $\phi(\tau)$ (coefficients T_i are defined as in [14]). Jones [35] introduced Wagner function's classical finite-state approximation and eight first-order differential equations by Edwards [36] were given as

$$\dot{\mathsf{v}} = \mathsf{A}(\mathsf{U})\mathsf{v} + \mathsf{g}(\mathsf{v},\mathsf{U}), \qquad \text{(8.18)}$$

where space state vector is given by $v^T = \{\xi, \alpha, \beta, u, \dot\xi, \dot\alpha, \dot\beta, \tilde{u}\}^T$, while $g(v, U)$ and $A(U)$ are non-linear and linear parts of motion equations, respectively.

The driving factor of energy harvesting is assumed to be the plunge motion and can be written as [37]

$$q - k_{ME}\xi + C_P v = 0, \qquad \text{(8.19)}$$

152 Piezoelectric Aeroelastic Energy Harvesting

where the electrode electric charge is represented by q, the inherent capacitance of PZT is C_P, the coupling electromechanical factor is given by k_{ME}, the plunge displacement is ξ, the potential between two nodes is v. In a closed-loop circuit, the external and internal resistances are parallel to each other and therefore the internal resistance is higher than the external resistance, so

$$-R\dot{q} = v. \tag{8.20}$$

The external resistance in the circuit is represented by R.

For single-layer PAEH, the charge collected on the electrode surface is evaluated by Lu et al. [37] approach as

$$q = \frac{dhe_{31}}{2}\delta\phi\xi - dL\epsilon_{33}\frac{v}{\delta}, \tag{8.21}$$

where δ represents the PZT layer thickness, d denotes the PZT patch width, the flexibility change of patch can be evaluated by approximating the wing's plunge mode with fixed-free beam bending mode represented by $\delta\phi$, ϵ_{33} is the dielectric constant, h is the beam thickness, L is the length of PZT over the beam, and e_{31} is the PZT constant in the 31 coupling direction, also

$$k_{ME} = \frac{dhe_{31}}{2}\delta\phi, \tag{8.22}$$

$$C_P = \frac{dL\epsilon_{33}}{\delta}. \tag{8.23}$$

Calculating output power by using Ohm's law gives

$$P = IV = I^2R, \tag{8.24}$$

$$\dot{q} = -\frac{1}{RC_P}q + \frac{1}{RC_P}k_{ME}\xi = I. \tag{8.25}$$

The overall output power of the piezoelectric aeroelastic energy harvester is expressed mathematically in Eq. (8.26) as

$$P = \frac{1}{RC_P^2}(q^2 + k_{ME}^2\xi^2 - 2qk_{ME}\xi). \tag{8.26}$$

The output power generated by the aeroelastic piezoelectric energy harvester is represented in Fig. 8.5.

Flutter-based aeroelastic energy harvesting 153

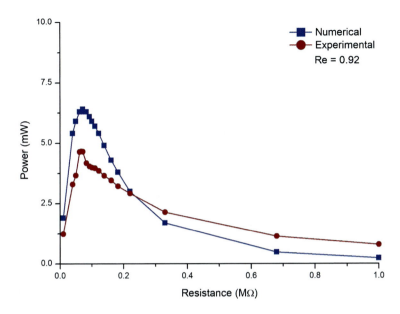

Figure 8.5 Output power of aeroelastic piezoelectric energy harvester; correlation factor of 0.92 between numerical and experimental results.

References

[1] C.A. Harvey, Adaptive control algorithm for flutter suppression, in: 1983 American Control Conference, IEEE, 1983, pp. 801–802.
[2] Abhijit Chakravarty, Dagfinn Gangsaas, John B. Moore, Adaptive flutter suppression in the presence of turbulence, in: 1985 American Control Conference, IEEE, 1985, pp. 751–756.
[3] A. Chakravarty, J.B. Moore, Aircraft flutter suppression via adaptive LQG control, in: 1986 American Control Conference, IEEE, 1986, pp. 488–493.
[4] Zhang JunHong, Han Jinglong, Zhang HongLiang, The flutter suppression of composite panel and the actuator placement optimization, in: 2010 2nd International Conference on Advanced Computer Control, vol. 4, IEEE, 2010, pp. 180–184.
[5] Naoki Kawai, Flutter control of wind tunnel model using a single element of piezoceramic actuator, in: Proceedings of the 24th International Congress of the Aeronautical Sciences, Yokohama, Japan, vol. 5, 2004, pp. 1–3.
[6] Michael W. Kehoe, et al., A Historical Overview of Flight Flutter Testing, vol. 4720, National Aeronautics and Space Administration, Office of Management ..., 1995.
[7] H. Ohta, A. Fujimori, P.N. Nikiforuk, M.M. Gupta, Active flutter suppression for two-dimensional airfoils, Journal of Guidance, Control, and Dynamics 12 (2) (1989) 188–194.
[8] Max Demenkov, From local to global stabilizability of aeroelastic oscillations, in: 2009 IEEE Control Applications (CCA) & Intelligent Control (ISIC), IEEE, 2009, pp. 131–135.
[9] Na Zhao, Dengqing Cao, Hongna Gao, Active flutter suppression for a 2D supersonic airfoil with nonlinear stiffness, in: 2010 3rd International Symposium on Systems and Control in Aeronautics and Astronautics, IEEE, 2010, pp. 493–497.

[10] WenHong Xing, Sahjendra N. Singh, Adaptive output feedback control of a nonlinear aeroelastic structure, Journal of Guidance, Control, and Dynamics 23 (6) (2000) 1109–1116.

[11] Shana D. Olds, Modeling and LQR control of a two-dimensional airfoil, PhD thesis, Virginia Tech, 1997.

[12] Michael S. Branicky, Multiple Lyapunov functions and other analysis tools for switched and hybrid systems, IEEE Transactions on Automatic Control 43 (4) (1998) 475–482.

[13] S.Y. Xu, T.W. Chen, Robust H_∞ control for uncertain stochastic systems with state delay, IEEE Transactions on Automatic Control 47 (12) (2002) 2089–2094.

[14] Theodore Theodorsen, W.H. Mutchler, General theory of aerodynamic instability and the mechanism of flutter, 1935.

[15] Glenn O. Thompson, Gerald J. Kass, Active flutter suppression – an emerging technology, Journal of Aircraft 9 (3) (1972) 230–235.

[16] Hassan Elahi, Marco Eugeni, Paolo Gaudenzi, Electromechanical degradation of piezoelectric patches, in: Analysis and Modelling of Advanced Structures and Smart Systems, Springer, 2018, pp. 35–44.

[17] H. Elahi, A. Israr, R.F. Swati, H.M. Khan, A. Tamoor, Stability of piezoelectric material for suspension applications, in: 2017 Fifth International Conference on Aerospace Science & Engineering (ICASE), IEEE, 2017, pp. 1–5.

[18] Vittorio Memmolo, Hassan Elahi, Marco Eugeni, Ernesto Monaco, Fabrizio Ricci, Michele Pasquali, Paolo Gaudenzi, Experimental and numerical investigation of PZT response in composite structures with variable degradation levels, Journal of Materials Engineering and Performance 28 (6) (2019) 3239–3246.

[19] Muhammad Usman Khan, Zubair Butt, Hassan Elahi, Waqas Asghar, Zulkarnain Abbas, Muhammad Shoaib, M. Anser Bashir, Deflection of coupled elasticity–electrostatic bimorph PVDF material: theoretical, FEM and experimental verification, Microsystem Technologies 25 (8) (2019) 3235–3242.

[20] Ahsan Ali, Riffat Asim Pasha, Hassan Elahi, Muhammad Abdullah Sheeraz, Saima Bibi, Zain Ul Hassan, Marco Eugeni, Paolo Gaudenzi, Investigation of deformation in bimorph piezoelectric actuator: analytical, numerical and experimental approach, Integrated Ferroelectrics 201 (1) (2019) 94–109.

[21] Mohsin Ali Marwat, Weigang Ma, Pengyuan Fan, Hassan Elahi, Chanatip Samart, Bo Nan, Hua Tan, David Salamon, Baohua Ye, Haibo Zhang, Ultrahigh energy density and thermal stability in sandwich-structured nanocomposites with dopamine@Ag@BaTiO$_3$, Energy Storage Materials 31 (2020) 492–504.

[22] Hassan Elahi, Zubair Butt, Marco Eugnei, Paolo Gaudenzi, Asif Israr, Effects of variable resistance on smart structures of cubic reconnaissance satellites in various thermal and frequency shocking conditions, Journal of Mechanical Science and Technology 31 (9) (2017) 4151–4157.

[23] Hassan Elahi, Marco Eugeni, Paolo Gaudenzi, Faisal Qayyum, Raees Fida Swati, Hayat Muhammad Khan, Response of piezoelectric materials on thermomechanical shocking and electrical shocking for aerospace applications, Microsystem Technologies 24 (9) (2018) 3791–3798.

[24] Hassan Elahi, Marco Eugeni, Paolo Gaudenzi, Madiha Gul, Raees Fida Swati, Piezoelectric thermo electromechanical energy harvester for reconnaissance satellite structure, Microsystem Technologies 25 (2) (2019) 665–672.

[25] Hassan Elahi, Khushboo Munir, Marco Eugeni, Muneeb Abrar, Asif Khan, Adeel Arshad, Paolo Gaudenzi, A review on applications of piezoelectric materials in aerospace industry, Integrated Ferroelectrics 211 (1) (2020) 25–44.

[26] Hassan Elahi, Marco Eugeni, Federico Fune, Luca Lampani, Franco Mastroddi, Giovanni Paolo Romano, Paolo Gaudenzi, Performance evaluation of a piezoelectric energy harvester based on flag-flutter, Micromachines 11 (10) (2020) 933.

[27] Marco Eugeni, Hassan Elahi, Federico Fune, Luca Lampani, Franco Mastroddi, Giovanni Paolo Romano, Paolo Gaudenzi, Numerical and experimental investigation of piezoelectric energy harvester based on flag-flutter, Aerospace Science and Technology 97 (2020) 105634.

[28] Marco Eugeni, Hassan Elahi, Federico Fune, Luca Lampani, Franco Mastroddi, Giovanni Paolo Romano, Paolo Gaudenzi, Experimental evaluation of piezoelectric energy harvester based on flag-flutter, in: Conference of the Italian Association of Theoretical and Applied Mechanics, Springer, 2019, pp. 807–816.

[29] Hassan Elahi, Marco Eugeni, Paolo Gaudenzi, Design and performance evaluation of a piezoelectric aeroelastic energy harvester based on the limit cycle oscillation phenomenon, Acta Astronautica 157 (2019) 233–240.

[30] Hassan Elahi, The investigation on structural health monitoring of aerospace structures via piezoelectric aeroelastic energy harvesting, Microsystem Technologies (2020) 1–9.

[31] Hassan Elahi, Marco Eugeni, Luca Lampani, Paolo Gaudenzi, Modeling and design of a piezoelectric nonlinear aeroelastic energy harvester, Integrated Ferroelectrics 211 (1) (2020) 132–151.

[32] Hassan Elahi, Khushboo Munir, Marco Eugeni, Paolo Gaudenzi, Reliability risk analysis for the aeroelastic piezoelectric energy harvesters, Integrated Ferroelectrics 212 (1) (2020) 156–169.

[33] Alper Erturk, Onur Bilgen, Matthieu Fontenille, Daniel J. Inman, Piezoelectric energy harvesting from macro-fiber composites with an application to morphing-wing aircrafts, in: Proceedings of the 19th International Conference on Adaptive Structures and Technologies, Citeseer, 2008, pp. 6–9.

[34] Franco Mastroddi, Daniele Dessi, Marco Eugeni, Pod analysis for free response of linear and nonlinear marginally stable aeroelastic dynamical systems, Journal of Fluids and Structures 33 (2012) 85–108.

[35] Robert T. Jones, The Unsteady Lift of a Wing of Finite Aspect Ratio, vol. 681, NACA, 1940.

[36] John W. Edwards, Holt Ashley, John V. Breakwell, Unsteady aerodynamic modeling for arbitrary motions, AIAA Journal 17 (4) (1979) 365–374.

[37] F. Lu, H.P. Lee, S.P. Lim, Modeling and analysis of micro piezoelectric power generators for micro-electromechanical-systems applications, Smart Materials and Structures 13 (1) (2003) 57.

CHAPTER 9

Limit cycle oscillations

Contents

9.1.	Introduction	157
9.2.	Non-linear aeroelastic system	160
9.3.	Non-linear aeroelastic systems's LCO	162
	9.3.1 Numerical modeling of LCOs	163
	9.3.2 Non-linear aeroelastic model	164
9.4.	Theoretical modeling of aeroelastic harvester	168
9.5.	Aerodynamic modeling of aeroelastic harvester	169
9.6.	Structural model	172
	9.6.1 Non-linear electroelastic equation of motion	173
	9.6.2 Aeroelectroelastic state space equations	175
References		176

9.1 Introduction

Once an aeroelastic system became unstable from the linear point of view the nonlinear mechanisms became important to determine the response of the system. In particular, some of this phenomena can be used for energy harvesting purpose because they are characterized by a persistent and self-induced motion. Among them, the LCO are most simple to be observed and used for energy harvesting purpose but of course their analysis is important also for aircraft design [1,2].

Linear models cannot simulate these non-linear aeroelastic problems accurately. Understanding, accurately predicting, and controlling physically critical terms, which dominate the non-linear aeroelastic phenomena, are important for the high performance of the new-generation aircraft.

In particular, LCOs were a persistent problem with many designs of fighter aircraft and were usually encountered in external store configurations. During the flight test of F-18 and F-16 aircraft, LCOs instabilities were experienced and this problem was discussed by Bunton et al. [3]. F-16 craft's extensive LCOs experiences were further elaborated by Denegri [4] through the analysis and flight tests. Flight-test data of aircraft's LCOs with various store configurations were discussed by Stearman et al. [5]. Beyond the characteristics of LCOs, a lot of non-linear characteristics are obvious. In the F-16 wing torsional stiffness, evidence of spring-hardening-type non-linearity was found by Denegri et al. [1] and Chen et al. [6]. At

Piezoelectric Aeroelastic Energy Harvesting
https://doi.org/10.1016/B978-0-12-823968-1.00020-9

Copyright © 2022 Elsevier Inc.
All rights reserved.

157

a high angle of attack, the LCO was observed at a Mach number as low as $M = 0.6$ and was found to be sensitive to the tip missile configuration. O'Neil et al. [7] did a comparison for the continuous non-linear stiffening behavior, between analytical predictions and experimental measurements in the pitch mode. LCOs are induced by the existence of torsion stiffness non-linearities, which are dependent on the non-linear parameters and velocity. An investigation of cubic stiffness non-linearities and the effects of hysteresis on the flutter characteristics was done by Woolston et al. [8].

Mission performance is restricted due to the store-induced LCO because acceleration which is induced by LCO causes loading for both the structure and pilot, which is unacceptable. The mechanism causing LCOs is not very well understood but it is possible to have some explanations using the studies conducted by the researchers, including aerodynamic and/or structural non-linearities.

Many experiments have shown that aeroelastic systems are essentially non-linear, which contributes to processes not adequately represented by linear representations. Most significantly, for the system exhibiting non-linearities, the LCO behavior occurs. This leads to velocity and non-linear parameter-dependent LCOs. The motion of the limit cycle consists of several components of higher harmonics in the case of continuous non-linearities, in addition to dominant flutter frequency.

Ground observation and flight-test measurement of F/A-18 and F-16 [1,6] provide motivation for further research. Observations from the flight and LCO measurements of a ground test on F/A-18, as got from the R. Yurkovich's panel discussion at the 41st AIAA Structural Dynamics and Materials Conference titled "Limit Cycle Oscillation and Related Nonlinear Phenomena in Aircraft" in 2020, are reviewed bellow:

1. Stiffness tests provide evidence for the non-linearity of hardening of spring in torsional stiffness of a wing, and it can be seen from the measurements that different behavior arises for the leading-edge down vs leading-edge up.

2. Data from accelerometers and the unsteady wing pressure shows harmonics that are twice as large as that of the LCO frequency, but the structural modes occur at this frequency.

3. The frequency of a pylon pitch is a little bit below the bending frequency, which is damping-sensitive. In some cases the bending of the lateral fuselage was included.

4. The mechanism of the linear flutter is antisymmetric and with wing bending, it involves pylon pitch coupling. For the critical stores and

their combinations, the linear analysis predicts flutter, but LCOs are found to occur at a speed lower than the speed predicted for the linear flutter.

5. There is sensitivity of LCO to the angle of attack (AOA), as AOA can rise positively or decrease accordingly; AOA occurs at high flight conditions.

6. With the Mach number less than 1, LCO occurs in the transonic speed range. LCOs occur at elevated AOA for numbers as low as $M = 0.6$.

7. LCOs are sensitive to the tip of the missile configuration. Changing the missile tip changes the occurrence of LCOs.

The observed results were identical to the behavior of LCOs for F-16 as mentioned by Denegri [4] and Chen et al. [6]. Similar to the LCO response of store configuration, an aeroelastic behavior was observed by Cole [9]. The presence of internal resonance was observed in the results. Nonlinearities of the system's couple of modes of motion result in internal resonance. It is not possible to predict the internal resonance with linear analysis [10,11]. In addition, the results obtained during the investigation of the wind-tunnel for aeroelastic behaviors indicate a similar response as that found in the LCO response of store configuration. During the wind tunnel test, an unexpected response was observed while comparing the predicted result with the linear theory based computational methods. Limit cycle flutter was found below that predicted by the linear analysis of the classical flutter. On the basis of the partial feedback linearization, a nonlinear controller was designed by Ko et al. [12], and as a result, the closed-loop stability is investigated further by Ko et al. [13]. The linearization of feedback depends critically on non-linearities' exact cancelation.

Response characteristics of dynamic systems were examined by Oh et al. [14] which was similar to Cole's [9] experiment; moreover, these experiments were performed without aerodynamic loads. According to the results, instability can be caused by two-to-one internal resonance because of the second bending mode's frequency of the first torsional mode. Results also show that there is a possibility to excite indirectly the antisymmetric vibration mode by a mechanism two-to-one internal resonance.

Non-linear behavior in aeroelastic systems was discussed by Denegri et al. [1] and their efforts were delineated to qualify using the linear analysis of the aeroelastic behavior. An examination of the store-induced LCOs was performed by Denegri (F-16) and Yurkovich (F-18), both made use of the traditional approaches and got limited success as in the velocity-damping

160   Piezoelectric Aeroelastic Energy Harvesting

diagram, hump modes (or soft crossing) were related to the LCOs which were store-induced.

9.2 Non-linear aeroelastic system

In a two-dimensional wing non-linear aeroelastic control problem, a wing is mounted on a flexible support which permits the two-degree-of-freedom motion as represented in Fig. 9.6. The plunge is represented by h and pitch variables are denoted by α. With these variables, the aeroelastic system is obtained as [15]

$$\begin{bmatrix} m_T & m_W x_\alpha b \\ m_W x_\alpha b & I_\alpha \end{bmatrix} \begin{Bmatrix} \ddot{h} \\ \ddot{\alpha} \end{Bmatrix} + \begin{bmatrix} c_h & 0 \\ 0 & c_\alpha \end{bmatrix} \begin{Bmatrix} \dot{h} \\ \dot{\alpha} \end{Bmatrix} + \begin{bmatrix} k_h & 0 \\ 0 & k_\alpha(\alpha) \end{bmatrix} \begin{Bmatrix} h \\ \alpha \end{Bmatrix} = \begin{Bmatrix} -L \\ M \end{Bmatrix},$$

(9.1)

where m denotes the wing's mass, and mass moment of inertia about the elastic axis is represented by I_α. Wing's mass is represented by m_W, whereas the total mass of support structure and the wing is denoted by m_T; x_α represents the dimensionless distance between the elastic axis and mass center, c_α and c_h are pitch structural damping and plunge coefficients, respectively. The aerodynamics lift is L and the moment is considered about the elastic axis. The structural stiffness is k_α for pitch motions and k_h for k_h. Various non-linear features can be incorporated into the design and the experiments associated with it. These forms include Coulomb damping, aerodynamic non-linear loads, higher-order kinematics, non-linear stiffness, etc. A non-linear torsional stiffness is approximated in the polynomial form as

$$k_\alpha(\alpha) = k_{\alpha_0} + k_{\alpha_1}\alpha + k_{\alpha_2}\alpha^2 + k_{\alpha_3}\alpha^3 + k_{\alpha_4}\alpha^4 + \cdots.$$

(9.2)

Non-linear cams are used in the experiment to realize the non-linear torsional stiffness, whereas in the preceding polynomial representation the actual coefficients are obtained from the displacement measurement and measurements of the load.

Many aerodynamic models are present for the representation of the moment loads and the unsteady lift. A quasi-steady aerodynamic model was used for the adaptive non-linear controller development [15,16]:

$$L = \rho U^2 b c_{l_\alpha} [\alpha + \frac{\dot{h}}{U} + (\frac{1}{2} - a)b(\frac{\dot{\alpha}}{U})] + \rho U^2 b c_{l_\beta} \beta,$$

(9.3)

$$M = \rho U^2 b^2 c_{m_\alpha} [\alpha + \frac{\dot{h}}{U} + (\frac{1}{2} - a)b(\frac{\dot{\alpha}}{U})] + \rho U^2 b^2 c_{m_\beta} \beta.$$

(9.4)

Isidori [17] and Slotine et al. [18] gave the theoretical derivation of the feedback linearization. The pitch variable is chosen as the output function,

$$y(x) = x_2 = \alpha. \tag{9.5}$$

It is supposed that the stabilization of the pitch motion is the prime purpose of the control system. A study of the zero dynamics of a problem of linearized partial feedback can help deduce the plunge motion stability. The output function's choice has a relative degree of 2 through the following equations:

$$y(x) = x_2, \qquad L_g y(x) = 0, \qquad L_f y(x) = x_4, \tag{9.6}$$
$$L_g L_f y(x) = g_4 U^2 \neq 0. \tag{9.7}$$

In Eq. (9.7), the Lie derivative of y in the direction f is represented by $L_y y$ and is given by

$$L_f y(x) = \sum_i \frac{\partial y}{\partial x_i} f_i. \tag{9.8}$$

The following state transformations are introduced:

$$x \mapsto \phi, \qquad \phi_1 = y(x) = x_2, \qquad \phi_2 = L_f y(x) = x_4, \tag{9.9}$$
$$\phi_3 = x_1, \qquad \phi_4 = -g_3 x_4 + g_4 x_3. \tag{9.10}$$

The system of equivalent transformation is obtained as

$$\dot{\phi}_1 = \phi_2, \qquad \dot{\phi}_2 = L_f^2 y(x) + L_g L_f y(x)\beta, \tag{9.11}$$
$$\dot{\phi}_3 = A_{32}\phi_2 + A_{34}\phi_4, \tag{9.12}$$
$$\dot{\phi}_4 = [g_3 P_U(\phi_1) - g_4 Q_U(\phi_1)]\phi_1 + A_{42}\phi_2 + A_{43}\phi_3 + A_{44}\phi_4, \tag{9.13}$$

where

$$L_f^2 y(x) = P_U(\phi_1)\phi_1 - [c_4 + c_3(\frac{g_3}{g_4})]\phi_2 - k_3\phi_3 - (\frac{c_3}{g_4})\phi_4, \tag{9.14}$$
$$P_U(\phi_1) = k_4 U^2 + q(\phi_1), \qquad Q_U(\phi_1) = k_2 U^2 + p(\phi_1), \tag{9.15}$$
$$A_{32} = \frac{g_3}{g_4}, \qquad A_{34} = \frac{1}{g_4}, \tag{9.16}$$
$$A_{42} = -c_1 g_3 - c_2 g_4 + c_3(\frac{g_3^2}{g_4}) + c_4 g_3, \qquad A_{43} = k_3 g_3 - k_1 g_4, \tag{9.17}$$

$$A_{44} = c_3\left(\frac{g_3}{g_4}\right) - c_1.\tag{9.18}$$

These equations are later transformed to the canonical form, which is known as amenable to non-linear geometric control. For all flow speeds, U, the given transforms and their inverses are defined [12]. Choosing the feedback control can lead to partial feedback linearization:

$$\beta = \frac{-L_f^2\gamma(x) + \nu}{L_g L_f \gamma(x)},\tag{9.19}$$

where control input is denoted by ν. When the ν structure is defined with the help of a linear control methodology, and one sets $\phi_1 = \phi_2 = 0$, it results in the zero dynamics system,

$$\begin{Bmatrix}\dot{\phi}_3 \\ \dot{\phi}_4\end{Bmatrix} = \begin{bmatrix} 0 & A_{34} \\ A_{43} & A_{44}\end{bmatrix}\begin{Bmatrix}\phi_3 \\ \phi_4\end{Bmatrix}.\tag{9.20}$$

The definition of zero dynamics is the internal dynamics of a partially linearized system. Hence the stability of the zero dynamics is important for the stability of the entire system. As it was seen, the set of linear equations represents the zero dynamics, therefore investigating the eigenvalues can help in checking the stability.

9.3 Non-linear aeroelastic systems's LCO

Non-linear LCO's numerical predictions are investigated and validated experimentally by a model system that was developed for the provision of direct measurements of non-linear aeroelastic responses [7]. Prescribed pitch and plunge motions are permitted by the support system for a mounted wing section. Carriage provides the plunge motion which is translated freely. On this carriage, a rotational carriage is mounted, which provides the pitch motion. It permits a large AOA but, on the amplitude of motion, there are some constraints for preventing the tunnel/model damage from the large LCOs amplitude.

In the test conditions and for the given the parameters, freedom is provided by the model support system. A pair of cams governs the apparatus structural stiffness response, they also provide the tailored non-linear or linear stiffness response. Spring's stiffness, and pretension, as well as cam's shape dictate the nature of the non-linearity. Stiffness characteristics, eccentricity, the shape of the wing, and moment of inertia can be modified easily for parametric investigation. Measurement of systems response is done with an accelerometer and optical encoder that are employed for the tracking of motion in each degree of freedom.

Elahi et al. [19] performed an experimental campaign to analyze the aeroelastic energy harvesting based on LCOs. In this campaign, the fiberglass flag attached with a piezoelectric patch as in Fig. 9.1 and Al patch as in Fig. 9.2 are subjected to axial flow in a subsonic wind tunnel. When the airflow is increased up to a certain point, i.e., flutter velocity; the harvester starts flapping and moves away from LCOs as represented in Fig. 9.1 [19].

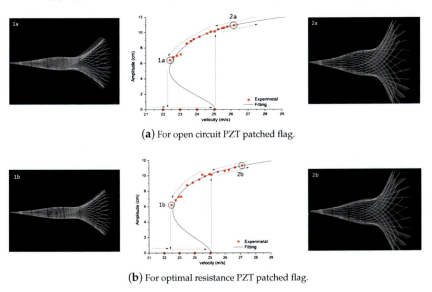

(a) For open circuit PZT patched flag.

(b) For optimal resistance PZT patched flag.

Figure 9.1 LCOs of an aeroelastic piezoelectric energy harvester via LCOs [19]; flutter occurs at 25 ms^{-1}.

9.3.1 Numerical modeling of LCOs

Elahi et al. performed a numerical campaign to investigate the deformation of the flags attached with Al and PZT patches after the occurrence of LCOs. This campaign is carried out in Matlab® and is presented in Fig. 9.3 [19]. Moreover, the authors validated the numerical results with experimental data which was in good agreement.

In order to have a clear picture of the LCOs, Beltramo et al. performed a study of flutter-based energy harvesters and calculated numerically the LCOs of the harvester. When the wing is subjected to the airflow having

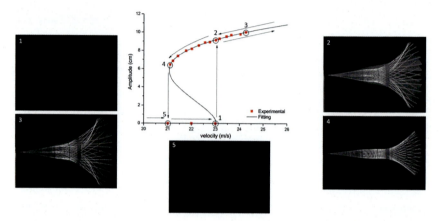

Figure 9.2 LCOs of an aeroelastic harvester attached with Al patch via LCOs [19]; flutter occurs at 23 ms^{-1}.

flow velocity equal to the flutter velocity of the wing, it experiences the phenomenon of LCOs, which is numerically represented in Fig. 9.4 [20].

The output variables obtained at the flow rate of a double-harvester range in terms of the unlimited separation distance $\frac{h}{c}$ during LCOs are elaborated in Fig. 9.5 [20].

9.3.2 Non-linear aeroelastic model

A model of a two-degree-of-freedom non-linear elastic system was designed by Essam et al. [21]. For the described aeroelastic system, non-linear equations of motion which describe the aeroelastic response are given as follows:

$$m_t \ddot{h} + [m_w x_\alpha b \cos(\alpha) - m_c r_c b \sin(\alpha)]\ddot{\alpha} + c_h \dot{h} + [-m_w x_\alpha b \sin(\alpha)$$
$$- m_c r_c b \cos(\alpha)]\dot{\alpha}^2 + k_h(h)h = -L(t) - \mu_h m_t g(\frac{|\dot{h}|}{\dot{h}}), \quad (9.21)$$

$$I_{EA}\ddot{\alpha} + [m_w x_\alpha b \cos(\alpha) - m_c r_c b \sin(\alpha)]\ddot{h} + c_\alpha \dot{\alpha} + k_\alpha(\alpha)\alpha$$
$$= M(t) - \mu_\alpha M_f(|\dot{\alpha}|/\dot{\alpha}), \quad (9.22)$$

where the wing and its support's total mass is defined by m_t and only wings mass is m_w; a is the location of the elastic axis which is nondimensionalized with respect to b, the half-chord length. The dimensionless distance between the cam's center of rotation and elastic axis is represented by r_c. The plunge and pitch viscous damping coefficients are c_h and c_α, respectively.

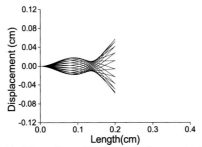
(a) 20 cm flag with Al patched numerical.

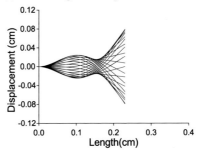
(b) 23 cm with Al patched flag numerical.

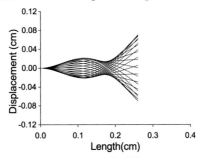
(c) 26 cm with Al patched flag numerical.

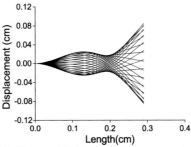
(d) 29 cm with Al patched flag numerical.

Figure 9.3 Numerical deformation comparison for Al and PZT patched flags, from 15 to 38 cm long [19].

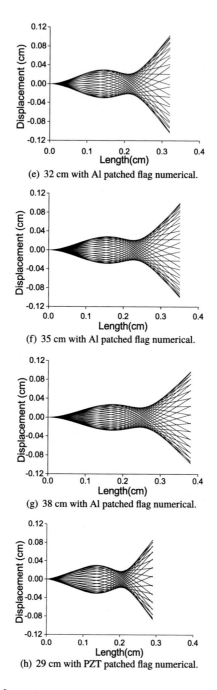

(e) 32 cm with Al patched flag numerical.

(f) 35 cm with Al patched flag numerical.

(g) 38 cm with Al patched flag numerical.

(h) 29 cm with PZT patched flag numerical.

Figure 9.3 (*continued*)

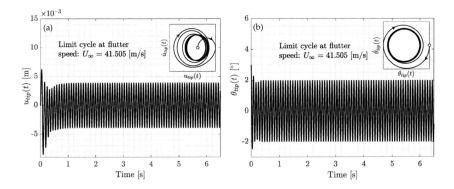

Figure 9.4 LCOs of the wing [20].

The mass moment of inertia I_{EA} about the elastic axis is the result of the contribution of cams, wing, and the offset masses and is given by

$$I_{EA} = I_{w_{cg}} + I_{C_{cg}} + m_c(r_c b)^2 + m_w(x_\alpha b)^2. \tag{9.23}$$

The plunge motion structural stiffness is denoted by k_h, and k_α is the pitch motion structural stiffness. In the equation above, the right-hand side is the Coulomb structural damping, where the frictional moment is represented by M_f, pitch motion's structural damping coefficient is given by μ_α, and plunge motion's structural damping coefficient is μ_h.

Moreover, L is the normal force and M is the pitching moment. It is possible to compute the aerodynamic forces by finding the solution for the full Navier–Stokes equations. Aerodynamic forces are referenced to an elastic axis and depend on the wing motion. In order to take into account the system's aerodynamic nonlinearities, Navier–Stokes equations must be solved. These non-linearities may occur because of the high angle of attack, viscosity effects, oscillating shocks, shock/boundary-layer interaction, shock-induced flow separation, shedding, and vorticity evolution.

Apart from the non-linearities of aerodynamics, other non-linear features such as non-linear stiffness, Coulomb structural damping, and high-order kinematics can influence the results. Aeroelastic module's non-linear stiffness is realized by the non-linear cams. An approximation by the fifth-order polynomial is assumed in the form of

$$k_\alpha(\alpha) = k_{\alpha_0} + k_{\alpha_1}\alpha + k_{\alpha_2}\alpha^2 + k_{\alpha_3}\alpha^3 + \cdots. \tag{9.24}$$

In this polynomial representation, the actual coefficients of the spring stiffness are obtained by loads' and displacement measurement.

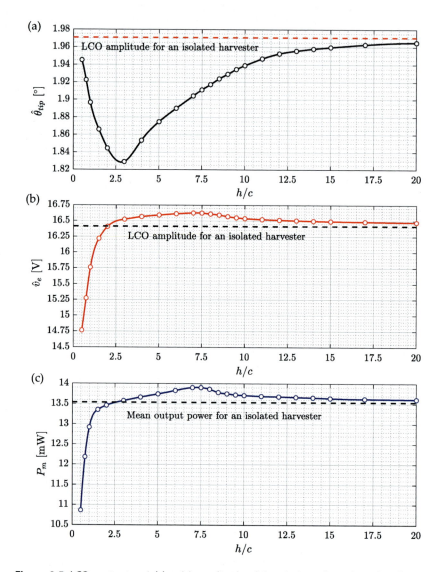

Figure 9.5 LCOs output variables: (a) amplitude of the pitch angle at the $\cap \theta_{tip}$, (b) amplitude of the voltage generated of one harvester $\cap v_{ei}$, (c) mean power generated P_m [20].

9.4 Theoretical modeling of aeroelastic harvester

The aeroelastic harvester is made up of a flat elastic cantilever panel connected with the bimorph configurations of piezoelectric laminates. The panel has thickness t, length L, and width b. At the beam's base, the fixed

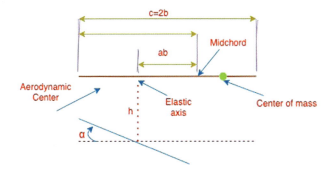

Figure 9.6 Aeroelastic structure model with pitch and plunge degrees of freedom.

boundary condition is achieved with the help of clamping with a leading edge to rigid airfoil of length L_0 and $b_{foil} \gg b$. Therefore the defined chord length for the harvester is $L + L_0$.

For the energy to be harvested from the self-excited LCO in the axial airflow, we designed a device. Inside the wind tunnel, there is a vertically mounted rigid airfoil to cantilever, a thin plate of aluminum that is parallel to the airflow direction. Large LCO amplitudes are attained when the aeroelastic damping occurs at the critical velocity of the wind speed. These large oscillation amplitudes cause strain from which extraction of the electric energy is possible using the piezoelectric transducers situated next to the base of the beam.

The transducer is made of a coupled piezoelectric bimorph with simple load impedance. We can tune the oscillation amplitude to the desired levels with the adjustment of the airspeed, whereas the change in the value of resistance of impedance load can adjust the electrical properties. The measurement of power is done with the help of recording the impedance and voltage across the load. The system's effective investigation in terms of electromechanical coupling, physical behavior, and potential of harvesting energy is possible using this device.

9.5 Aerodynamic modeling of aeroelastic harvester

The atmosphere offers energy. Different mechanisms can be used for processing and harvesting energy. To achieve energy collection from the ambient atmosphere, various instruments and techniques have been developed and proposed. These techniques include using piezoelectric [22–24], magnetoelectric [25,26], thermoelectric generators [27], and thermoacoustic devices [28–32].

Technologies for wind energy harvesting are well known. However, due to the high tip speed, the wind turbine blades produce loud aero/hydrodynamic noise [26]. And with a very low Reynolds number, efficiency decreases due to laminar separation [33]. Yet, increasing energy demand for small electronic equipment, such as remote sensors, makes the use of airflow kinetic energy essential.

Li et al. [34] have carried out a series of experiments to harness ambient wind energy using a piezoelectric generator. The cross-flow configuration has been shown to increase performance. Kwuimy et al. [23] carried out a theoretical study of the voltage output of the piezoelectric generator. In conditions near the stochastic resonances region, it has been shown that the harvester performs more effectively.

Non-linear limit cycle oscillations were observed experimentally by Erturk et al. [35], when the harvester response was analyzed under a uniform air flow. Kwon [36] determined the minimum airflow needed to trigger the LCOs, under axial airflow for a T-shaped piezoelectric cantilevered beam.

Vibration-based piezoenergy-harvesting systems efficiently operate in the narrow frequency bandwidth, where the fundamental frequency and the excitation frequency are closer to each other. The output energy can be reduced because of a small variation in the excitation frequency, which can also make the harvester work inefficiently [23]. This limits the vibration-based energy harvester application [37,38]. Airflow has the advantage that it is a constant source of kinetic energy. However, few designs and applications are documented in the field of aeroelastic–piezoelectric energy harvesting.

In order to have a full understanding of the system, it is essential to have an understanding of the behavior, as well as of intrinsic dynamics to the points where there is an accurate prediction of the experimental behavior. Formation of the basic theoretical concept of the involved system's aerodynamics is the first step. The considered aerodynamic model is an unsteady, linear, three-dimensional vortex–lattice framework for inviscid, irrotational, and incompressible flows. Tang et al. [39] used a 2D aerodynamic kernel function for the calculation of aerodynamic forces on system's discretized elements. In aerodynamic vortex–lattice discretization, wake is divided into ($kmm-km$) segments of length dx in the chord direction, the chord length is divided into km segments of length dx in the chord direction, and the panel width is divided into kn segments of length dy in the span direction. In the analysis, 60 is the value of km, 120 is taken for kmm, and 5 is for kn. Vortex horseshoe points are placed at each element's quarter chord, while the points of collocation are placed at each element's three-quarter chord. Collocation points need the matching of total induced velocity to the panel's unsteady motion. This leads to the following relationship:

$$W_i^{t+1} = \sum_j^{k_{mm}} K_{ij}\Gamma_j^{t+1}, \qquad i = 1,\ldots,k_m, \tag{9.25}$$

where at time step $t+1$ the induced velocity is represented by W_i^{t+1} and i is the collocation point, K_{ij} is the aerodynamic kernel, and the jth vortex strength is given by Γ_j. The aerodynamic kernel is

$$K_{ij} = \frac{-1}{4\pi(y_i - y_{ja})}[1 + \frac{\sqrt{(x_i - x_{ja})^2}\sqrt{(y_i - y_{ja})^2}}{x_i - x_{ja}}]$$
$$+ \frac{1}{4\pi(y_i - y_{jb})}[1 + \frac{\sqrt{(x_i - x_{ja})^2}\sqrt{(y_i - y_{jb})^2}}{x_i - x_{ja}}], \tag{9.26}$$

where the ith collocation point was denoted by the subscript i and subscripts jb and ja denote the jth pair location of trailing vortex lines in the chord direction [40]. The influence coefficients are calculated using the kernel functions on every panel's element with respect to every other panel point. A combination of the discretized vortex strengths with influence coefficients forms the aerodynamic matrix equation, which relates the vortex strengths to velocities induced throughout the aerodynamic system:

$$[A]\Big[\Gamma\Big]^{t+1} + [B]\Big[\Gamma\Big]^{t} = \Big[W\Big]^{t+1}, \tag{9.27}$$

where the given timestep is represented by the superscript t, and $t+1$ is for the next timestep, matrices [A] and [B] are composed of influence coefficients that govern the mutual interactions of discrete vortex elements.

Assuming a linear relationship between velocity $\dot{q}$, modal panel displacement q, and the induced velocity at collocation point as can be seen from Eq. (9.25), the aerodynamic forces can be indirectly related to the out-of-plane deflection by

$$\{W\} = [E]\{\theta\}, \tag{9.28}$$

where the panel coordinate vector is $\{\theta\} = [E]\{\theta\}$ [39] in the prediction of the generated power from self-excited panel oscillations' amplitude of the out-of-plane deflection. The matrix form of the equation governing

172 Piezoelectric Aeroelastic Energy Harvesting

system's aerodynamics can be written by combining Eqs. (9.27) and (9.28) as

$$[A]\{\Gamma\}^{t+1} + [B]\{\Gamma\}^{t} - [E]\{\theta\}^{t+1} = 0. \tag{9.29}$$

It is essential that for each point on the panel, a pressure distribution is defined so that it is possible to derive the mathematical equation for the generalized force on each element. Following Tang et al. [39] derivation, the expression for the normalized pressure distribution can be derived as

$$\bar{\Delta p}_j = \frac{c}{dx}\left[\frac{(\Gamma_j^{t+1} + \Gamma_j^{t})}{2} \sum_{j}^{i}(\Gamma_i^{t+1} - \Gamma_i^{t})\right], \tag{9.30}$$

where the normalized pressure is given by $\bar{\Delta p}_j$ and the chord length is denoted by c. At every Q_i point, the generalized aerodynamic force can be expressed as

$$Q_i = \rho_\infty U^2 \int_0^c \bar{\Delta p}\Phi_i dx, \tag{9.31}$$

where the downstream fluid density is denoted by ρ_∞, and the fluid velocity is given by U. In the above equation, Φ_i denotes the ith downwash mode function,

$$\Phi_i = \begin{cases} 0 & \text{for } x \leq L_0, \\ \phi_i & \text{for } L + L_0 \geq x \geq L_0, \end{cases} \tag{9.32}$$

where the ith structural bending mode function of the 2D cantilevered flat plate is denoted by ϕ_i.

It is presumed that there is a negligible effect on aeroelastic system dynamics of the modal mass and stiffness because of piezoceramic laminates. Stanton et al. [41] describe a linear heterogeneous beam's dynamic response that is different from continuous beams. Erturk et al. [42] further studied this issue. Tang et al. [39] used $\alpha = 0.992$ in their computation as the vortex relaxation factor.

9.6 Structural model

Flapping flag's kinetic energy T is written as

$$T = \frac{1}{2}m_p \int_0^L (\dot{u}^2 + \dot{w}^2)dx, \tag{9.33}$$

where the in-plane deflection is denoted by u, the out-of-plane deflection is represented by w, and the plate's mass per unit length is given by m_p. The potential energy V is

$$V = \frac{1}{2} \int_0^L D\psi''^2 dx, \qquad (9.34)$$

where the panel's flexural rigidity is represented by D and the non-linear curvature formulation ψ'' is as utilized by Semler et al. [43]. At the point where the whole system energy is kinetic and conserves energy (i.e., when the beam is restored to the undeflected state), the power P delivered to the system can be approximated as

$$P = \dot{T} + \dot{V} = \dot{T}_{\max}. \qquad (9.35)$$

This equation assumes that the necessary power for maintaining the transfer of energy between the kinetic (K.E.) and potential (P.E.) energy forms at the defined rate by the time required for the system movement from zero K.E. to zero P.E. state, which by flow gets delivered to the beam. Tang et al. [44] discussed the energy transfer formula in an axial flow of cantilevered plates. Practically, the extraction of a power measurement delivered to the system can be approximated using the time derivatives by the multiplication of the system oscillation frequency in radians per second, ω. The derivative concept is valid by assuming the simple harmonic at the dominant aeroelastic mode. Therefore the expression of the fluid power delivered to the beam is

$$P = \dot{T}_{\max} \approx \frac{1}{2} m_p \omega \int_0^L (w\omega)^2 dx. \qquad (9.36)$$

For the estimation of the system's efficiency, the above expression is found more useful.

9.6.1 Non-linear electroelastic equation of motion

Using Hamiltons' principle, it is possible to derive the equation of motion in which a formulation of kinetic and potential energy applied to the oscillating plate is used [39,41]. We use the model of an electroelastic non-linear beam by Stanton et al. [41] and the non-linear uniform aeroelastic beam model of Tang et al. [39]. If a non-linear beam undergoes a large amplitude motion then each analysis includes the application of condition of inextensibility along the bending neutral axis [43,45]. The author of [39,41]

174 Piezoelectric Aeroelastic Energy Harvesting

assumed a small out-of-plane deflection and that the mechanical bending stress in both the thickness and span directions is negligible. If the results of Stanton et al. of the cantilevered beam undergoing arbitrary forcing augmented with a piezoelectric bimorph are combined with the results of Tang et al. for the uniform cantilevered plate with an electromechanical model in the axial flow then a system consisting of the coupled non-linear ordinary differential equations can be obtained as

$$M_{ii}\ddot{q} + \sum_n \sum_r \sum_s M_{inrs} q_n q_r \ddot{q}_s + \omega_i^2 M_{ii} q_i + F_K + F_M + \Theta_i v = Q_i, \qquad (9.37)$$

$$C\dot{v} + \frac{1}{R}v + \Theta_i \dot{q}_i = 0, \qquad (9.38)$$

where q_i denotes the attributable modal panel displacement to the ith structural mode normalized by L, the electric load across absolute voltage is v, element's mass matrix is given by M_{ii}, F_K is the non-linear force on cantilever plate, from the resultant curvature F_M, and the non-linear inertial forces are $\sum_n \sum_r \sum_s M_{inrs} q_n q_r \ddot{q}_s$, while ω_i is the natural frequency of the panel's model. In the electrical network equation, Θ is the modal electromechanical coupling vector, the effective capacitance of the in series connected piezoelectric laminates is C, and the electrical system's effective resistance is R. Suitable terms for damping, stiffness, and inertia are provided as

$$M_{ii} = \int_0^1 m_p \phi_i^2 dx, \qquad (9.39)$$

$$F_M = \sum_n \sum_r \sum_s M_{inrs} q_n \dot{q}_r \dot{q}_s, \qquad (9.40)$$

$$F_K = \sum_n \sum_r \sum_s K_{inrs} q_n q_r q_s, \qquad (9.41)$$

$$K_{inrs} = \int_0^1 D\phi_i [\phi_n'''' \phi_r' \phi_s' + 4\phi_n'' \phi_r'' \phi_s'' + \phi_n'' \phi_r'' \phi_s''] dx, \qquad (9.42)$$

$$M_{inrs} = \int_0^1 m_p \phi_i \phi_n' (\int_0^x \phi_r' \phi_s' dx) dx - \int_0^1 m_p \phi_i \phi_n'' (\int_x^1 \int_0^x \phi_r' + \phi_s' dx dx) dx, \qquad (9.43)$$

where flexural rigidity is given by D, and mass per unit chord length of plate is m_p [39]. The piezoelectric bimorph's expressions of equivalent ca-

Limit cycle oscillations **175**

pacitance and electromechanical coupling vector are

$$C = \frac{\epsilon_{33} b_p L_p}{2h_p},\tag{9.44}$$

$$\theta_i = \frac{1}{2} e_{31} b_p (h_p + h) \phi_i'(L_p),\tag{9.45}$$

where the permittivity is denoted by ϵ_{33}, the piezoelectric material's electromechanical coupling factor is represented by e_{31}, the piezoelectric section's width is denoted by b_p, the thickness of the piezoelectric transducer is h_p, the piezoelectric segment's length is L_p, and the plate thickness h [41].

9.6.2 Aeroelectroelastic state space equations

The transformation of Eqs. (9.37) and (9.38) into a single discrete equation in the state space for the flutter analysis using linear eigenvalue methods can yield the numerical solution for discrete time histories $v(t)$ and $q(t)$. Writing the governing equations in the state space form is the first transformation step:

$$\begin{bmatrix} M_{ii} & 0 & 0 & 0 \\ 0 & 1 & 0 & 0 \\ 0 & 0 & 0 & 0 \\ 0 & 0 & 0 & 1 \end{bmatrix} \begin{Bmatrix} \ddot{q} \\ \dot{q} \\ \ddot{v} \\ \dot{v} \end{Bmatrix} + \begin{bmatrix} 0 & \omega^2 M_{ii} & 0 & -\Theta_i \\ -1 & 0 & 0 & 0 \\ \Theta_j & 0 & C & \frac{1}{R} \\ 0 & 0 & -1 & 0 \end{bmatrix} \begin{Bmatrix} \dot{q} \\ q \\ \dot{v} \\ v \end{Bmatrix} = \begin{Bmatrix} Q_i - F_N \\ 0 \\ 0 \\ 0 \end{Bmatrix},\tag{9.46}$$

where the normalized nonlinear force F_N is defined as

$$F_N = \frac{F_K}{\tau^2} + \frac{F_M}{m_p L^2}.\tag{9.47}$$

Eq. (9.46) can be discretized by application of temporal differencing schemes to the terms involving v and q, taking the time increment as δt and general coordinates as x:

$$x_i = \frac{x^{t+1} + x^t}{2},\tag{9.48}$$

$$\dot{x}_i = \frac{x^{t+1} - x^t}{\Delta t},\tag{9.49}$$

$$\ddot{x}_i = \frac{\dot{x}^{t+1} - \dot{x}^t}{\Delta t}.\tag{9.50}$$

176 Piezoelectric Aeroelastic Energy Harvesting

If the discretized forms of both voltage vector $\{v\}$ and panel response $\{\theta\}$ are substituted then solving subsequently two new coefficient matrices allows rewritting Eq. (9.46) as

$$
\begin{bmatrix} D_2 & H_2 \\ J_2 & K_2 \end{bmatrix} \begin{Bmatrix} \theta \\ v \end{Bmatrix}^{t+1} + \begin{bmatrix} D_1 & H_1 \\ J_1 & K_1 \end{bmatrix} \begin{Bmatrix} \theta & v \end{Bmatrix}^{t} = \begin{Bmatrix} Q_i - F_N \\ 0 \end{Bmatrix}^{t+\frac{1}{2}}, \tag{9.51}
$$

where $[H_1]$, $[H_2]$, $[D_1]$, $[D_2]$, $[K_1]$, $[K_2]$, $[J_1]$, and $[J_2]$ are compact versions of four quadrants of new coefficient matrices. The aeroelectroelastic system's full state representation can be obtained by integrating the earlier derived aerodynamics into the structural model. In accomplishing this, the first step is to rewrite the aerodynamic force Q_i, in temporally discretized form, from the vortex–lattice formulation,

$$
- Q_i = [C_2]\{\Gamma\}^{t+1} + [C_1]\{\Gamma\}^{t}. \tag{9.52}
$$

This leads to the augmented form given by

$$
\begin{bmatrix} 0 & 0 & 0 \\ C_2 & D_2 & H_2 \\ 0 & J_2 & K_2 \end{bmatrix} \begin{Bmatrix} \Gamma \\ \theta \\ v \end{Bmatrix}^{t+1} + \begin{bmatrix} 0 & 0 & 0 \\ C_1 & D_1 & H_1 \\ 0 & J_1 & K_1 \end{bmatrix} \begin{Bmatrix} \Gamma \\ \theta \\ v \end{Bmatrix}^{t} = \begin{Bmatrix} 0 \\ -F_N \\ 0 \end{Bmatrix}^{t+\frac{1}{2}}.
$$

Now if the trivial first row is replaced with the aerodynamic matrix equality then this yields the aeroelectroelastic state-space model as

$$
\begin{bmatrix} A & -E & 0 \\ C_2 & D_2 & H_2 \\ 0 & J_2 & K_2 \end{bmatrix} \begin{Bmatrix} \Gamma \\ \theta \\ v \end{Bmatrix}^{t+1} + \begin{bmatrix} B & 0 & 0 \\ C_1 & D_1 & H_1 \\ 0 & J_1 & K_1 \end{bmatrix} \begin{Bmatrix} \Gamma \\ \theta \\ v \end{Bmatrix}^{t} = \begin{Bmatrix} 0 \\ -F_N \\ 0 \end{Bmatrix}^{t+\frac{1}{2}}.
$$

The latter equation completely incorporates the electrical, mechanical, and aerodynamic aspects into a single nonlinear aeroelectroelastic state-space expression of the flapping flag energy harvester.

References

[1] Charles Denegri Jr., Malcolm Cutchins, Charles Denegri Jr., Malcolm Cutchins, Evaluation of classical flutter analysis for the prediction of limit cycle oscillations, in: 38th Structures, Structural Dynamics, and Materials Conference, 1997, p. 1021.
[2] Gautam Shah, Wind tunnel investigation of aerodynamic and tail buffet characteristics of leading-edge extension modifications to the F/A-18, in: 18th Atmospheric Flight Mechanics Conference, 1991, p. 2889.

[3] Robert W. Bunton, Charles M. Denegri Jr., Limit cycle oscillation characteristics of fighter aircraft, Journal of Aircraft 37 (5) (2000) 916–918.

[4] Charles M. Denegri Jr., Limit cycle oscillation flight test results of a fighter with external stores, Journal of Aircraft 37 (5) (2000) 761–769.

[5] Ronald O. Stearman, E.J. Powers, Jason Schwartz, Rudy Yurkovich, Aeroelastic system identification of advanced technology aircraft through higher order signal processing, in: Proceedings of the 9th International Modal Analysis Conference (IMAC), vol. 2, Florence, Italy, 1991, pp. 1607–1616.

[6] P.C. Chen, D. Sarhaddi, D. Liu, Limit-cycle oscillation studies of a fighter with external stores, in: 39th AIAA/ASME/ASCE/AHS/ASC Structures, Structural Dynamics, and Materials Conference and Exhibit, 1998, p. 1727.

[7] Todd O'Neil, Thomas W. Strganac, Aeroelastic response of a rigid wing supported by nonlinear springs, Journal of Aircraft 35 (4) (1998) 616–622.

[8] Donald S. Woolston, Harry L. Runyan, Robert E. Andrews, An investigation of effects of certain types of structural nonlinearities on wing and control surface flutter, Journal of the Aeronautical Sciences 24 (1) (1957) 57–63.

[9] Stanley Cole, Effects of spoiler surfaces on the aeroelastic behavior of a low-aspect-ratio rectangular wing, in: 31st Structural Dynamics and Materials Conference, 1990, p. 981.

[10] A.H. Nayfeh, D.T. Mook, Nonlinear Oscillations, Wiley, New York, 1979; L.D. Landau, E.M. Lifshitz, Mechanics, Pergamon, New York, 1976, pp. 338–348.

[11] Ali H. Nayfeh, B. Balachandran, Modal interactions in dynamical and structural systems, 1989.

[12] Jeonghwan Ko, Andrew J. Kurdila, Thomas W. Strganac, Nonlinear control of a prototypical wing section with torsional nonlinearity, Journal of Guidance, Control, and Dynamics 20 (6) (1997) 1181–1189.

[13] Jeonghwan Ko, Thomas W. Strganac, Andrew J. Kurdila, Stability and control of a structurally nonlinear aeroelastic system, Journal of Guidance, Control, and Dynamics 21 (5) (1998) 718–725.

[14] Kyoyul Oh, Ali H. Nayfeh, Dean T. Mook, Modal interactions in the forced vibration of a cantilever metallic plate, American Society of Mechanical Engineers, Applied Mechanics Division, AMD 192 (1994) 237–247.

[15] Yuan-cheng Fung, An Introduction to the Theory of Aeroelasticity, Wiley, 1955.

[16] Thomas W. Strganac, Jeonghwan Ko, David E. Thompson, Andrew J. Kurdila, Identification and control of limit cycle oscillations in aeroelastic systems, Journal of Guidance, Control, and Dynamics 23 (6) (2000) 1127–1133.

[17] Alberto Isidori, Nonlinear Control Systems, Springer Science & Business Media, 2013.

[18] Jean-Jacques E. Slotine, Weiping Li, et al., Applied Nonlinear Control, vol. 199, Prentice Hall, Englewood Cliffs, NJ, 1991.

[19] Hassan Elahi, Marco Eugeni, Federico Fune, Luca Lampani, Franco Mastroddi, Giovanni Paolo Romano, Paolo Gaudenzi, Performance evaluation of a piezoelectric energy harvester based on flag-flutter, Micromachines 11 (10) (2020) 933.

[20] Emmanuel Beltramo, Martín E. Pérez Segura, Bruno A. Roccia, Marcelo F. Valdez, Marcos L. Verstraete, Sergio Preidikman, Constructive aerodynamic interference in a network of weakly coupled flutter-based energy harvesters, Aerospace 7 (12) (2020) 167.

[21] Essam F. Sheta, Vincent J. Harrand, David E. Thompson, Thomas W. Strganac, Computational and experimental investigation of limit cycle oscillations of nonlinear aeroelastic systems, Journal of Aircraft 39 (1) (2002) 133–141.

[22] Huicong Liu, Songsong Zhang, Ramprakash Kathiresan, Takeshi Kobayashi, Chengkuo Lee, Development of piezoelectric microcantilever flow sensor with wind-driven energy harvesting capability, Applied Physics Letters 100 (22) (2012) 223905.

[23] C.A. Kitio Kwuimy, G. Litak, M. Borowiec, C. Nataraj, Performance of a piezoelectric energy harvester driven by air flow, Applied Physics Letters 100 (2) (2012) 024103.

[24] D. St. Clair, A. Bibo, V.R. Sennakesavababu, M.F. Daqaq, G. Li, A scalable concept for micropower generation using flow-induced self-excited oscillations, Applied Physics Letters 96 (14) (2010) 144103.

[25] Pol Grasland-Mongrain, Jean-Martial Mari, Bruno Gilles, Jean-Yves Chapelon, Cyril Lafon, Electromagnetic hydrophone with tomographic system for absolute velocity field mapping, Applied Physics Letters 100 (24) (2012) 243502.

[26] Dan Zhao, Joe Khoo, Rainwater-and air-driven 40 mm bladeless electromagnetic energy harvester, Applied Physics Letters 103 (3) (2013) 033904.

[27] Rama Venkatasubramanian, Edward Siivola, Thomas Colpitts, Brooks O'Quinn, Thin-film thermoelectric devices with high room-temperature figures of merit, Nature 413 (6856) (2001) 597–602.

[28] Dan Zhao, Waste thermal energy harvesting from a convection-driven Rijke–Zhao thermo-acoustic-piezo system, Energy Conversion and Management 66 (2013) 87–97.

[29] Dan Zhao, Z.H. Chow, Thermoacoustic instability of a laminar premixed flame in Rijke tube with a hydrodynamic region, Journal of Sound and Vibration 332 (14) (2013) 3419–3437.

[30] J. Smoker, M. Nouh, O. Aldraihem, A. Baz, Energy harvesting from a standing wave thermoacoustic-piezoelectric resonator, Journal of Applied Physics 111 (10) (2012) 104901.

[31] Dan Zhao, Y. Chew, Energy harvesting from a convection-driven Rijke–Zhao thermoacoustic engine, Journal of Applied Physics 112 (11) (2012) 114507.

[32] Dan Zhao, Chenzhen Ji, C. Teo, Shihuai Li, Performance of small-scale bladeless electromagnetic energy harvesters driven by water or air, Energy 74 (2014) 99–108.

[33] Mohamed A. Sayed, Hamdy A. Kandil, Ahmed Shaltot, Aerodynamic analysis of different wind-turbine-blade profiles using finite-volume method, Energy Conversion and Management 64 (2012) 541–550.

[34] Shuguang Li, Jianping Yuan, Hod Lipson, Ambient wind energy harvesting using cross-flow fluttering, 2011.

[35] A.F. Arrieta, P. Hagedorn, Alper Erturk, D.J. Inman, A piezoelectric bistable plate for nonlinear broadband energy harvesting, Applied Physics Letters 97 (10) (2010) 104102.

[36] Soon-Duck Kwon, A t-shaped piezoelectric cantilever for fluid energy harvesting, Applied physics letters 97 (16) (2010) 164102.

[37] C. Boragno, R. Festa, A. Mazzino, Elastically bounded flapping wing for energy harvesting, Applied Physics Letters 100 (25) (2012) 253906.

[38] Michael S. Howe, Michael S. Howe, Acoustics of Fluid–Structure Interactions, Cambridge University Press, 1998.

[39] D.M. Tang, H. Yamamoto, E.H. Dowell, Flutter and limit cycle oscillations of two-dimensional panels in three-dimensional axial flow, Journal of Fluids and Structures 17 (2) (2003) 225–242.

[40] Joseph Katz, Allen Plotkin, Low-Speed Aerodynamics, vol. 13, Cambridge University Press, 2001.

[41] S.C. Stanton, B.P. Mann, Nonlinear electromechanical dynamics of piezoelectric inertial generators: modeling, analysis, and experiment, Nonlinear Dynamics (2010).

[42] Alper Erturk, Pablo A. Tarazaga, Justin R. Farmer, Daniel J. Inman, Effect of strain nodes and electrode configuration on piezoelectric energy harvesting from cantilevered beams, Journal of Vibration and Acoustics 131 (1) (2009).

[43] Christian Semler, Guang Xuan Li, M.P. Païdoussis, The non-linear equations of motion of pipes conveying fluid, Journal of Sound and Vibration 169 (5) (1994) 577–599.

[44] Liaosha Tang, Michael P. Païdoussis, Jin Jiang, Cantilevered flexible plates in axial flow: energy transfer and the concept of flutter-mill, Journal of Sound and Vibration 326 (1–2) (2009) 263–276.

[45] M.R.M. Crespo da Silva, C.C. Glynn, Nonlinear flexural–flexural–torsional dynamics of inextensional beams. I. Equations of motion, Journal of Structural Mechanics 6 (4) (1978) 437–448.

CHAPTER 10

Vortex-induced vibrations based aeroelastic energy harvesting

Contents

10.1. Introduction	181
10.2. Simple numerical example for a VIV aeroelastic energy harvester	183
10.2.1 Existence of a "critical mass"	184
10.3. Vortex-induced vibrations in circular cylinders	185
10.4. Energy harvesting form the PZT based on VIV	185
10.5. VIV-based energy harvesters	189
10.5.1 Circular cylinder based VIV energy harvesters	190
10.5.2 Energy harvester with turbulent flow	190
10.5.3 Usage of poled and electrode flexible ceramic cylinder for power harvesting	191
10.5.4 Energy harvesting from a rigid, elastically mounted spherical cylinder	191
10.5.5 Energy harvester composed of aluminum cantilevered shim	191
10.5.6 Energy harvester based on electromagnetic induction	192
10.5.7 Vortex-induced vibration aquatic clean energy	192
10.5.8 Passive turbulence control	192
10.5.9 Harvester's performance enhancement	192
10.5.10 Harvesting energy from rigid circular cylinder	194
References	195

10.1 Introduction

Many engineering fields are concerned with vortex-induced vibration (VIV) of systems. VIV's functional value is addressed in a vast number of basic studies, many of which were thoroughly examined in the reviews [1–4], books [5], and book chapters [6–8]. We are particularly interested in oscillating the rigid cylinder mounted elastically, forced movements of these systems, the two-degree-of-freedom bodies, as well as the cantilevers dynamics, pivoted tubes, cords, and tethered bodies. We will use an elastically–mounted cylinder as a model for such vibration systems, which is restrained so it is transversely moved to flow. When the flow rate U increases, the vortex forming frequency f_V reaches a state where it is almost equal to the body's normal frequency f_N, so that the wake vortex's unsteady pressure causes the body to respond. Body movement may activate such wake patterns, regardless of whether body movement is regulated or is

Piezoelectric Aeroelastic Energy Harvesting
https://doi.org/10.1016/B978-0-12-823968-1.00021-0

Copyright © 2022 Elsevier Inc.
All rights reserved.

181

free to vibrate because of the fluid forces. Williamson et al. [9] researched the forced sinusoidal trajectory vortex wake patterns over a large variety of wavelengths (l/D up to 15.0) and amplitudes (A/D up to 5.0). A collection of various regimes for vortex wake modes was described, in the ($l/D, A/D$) plane, where for each mode a descriptive terminology is added. Every periodic vortex wake pattern consists of vortex (P) pairs and single vortex (S), giving patterns such as 2P, 2S, and P+S modes, which are the main modes near the fundamental lock in the field. Many researchers are working on energy harvesting techniques by using piezoelectric materials [10–15].

Modes 2P and P+S were observed in controlled vibration experiments in-line with the flow [16,17], in addition to being transverse to the flow [18]. The P+S mode has also been observed in popular smoke visualizations. These controlled vibration modes are significant because they provide regime maps in which such free vibration branches can be observed. It is also important to note that forced vibration can result in different vortex modes, such as a P+S mode, which cannot do the excitation of the body towards free vibration [19–22]. Essentially, with the transfer of energy or when the work is done on the body by the fluid, over a positive cycle a nominal periodic vibration occurs, where the side force's phase is induced related to the motion of the body, influencing the net energy transfer to a greater extent. It has a connection to the timing of the dynamics of the vortex. The interesting feedback between vortex and body movements is also an issue with VIV.

Also in the simple case of an elastically assembled cylinder, there are several fundamental questions that arise from the state-of-the-art:

1. For extremely low mass and damping, what might be the greatest possible amplitude for a cylinder undergoing VIV?
2. Under what circumstances the peak-amplitude data collapses due to the implementation of the classical "mass damping parameter"?
3. For the peak amplitude versus mass ratio, what should be the functional shape?
4. What kind of bifurcation experience the aeroelastic system representing the VIV harvester?
5. Which is the relationship between the wake vortexes structures and aeroelastic dynamics for the VIV harvester and thus, with the gathered energy?
6. Is it possible to determine a general framework for the design of a VIV system?

7. To what degree the present state-of-the-art results about simople VIV aeroelastic harvesters can be applicable to more complex ones?
8. Since almost all of the VIV experiments have low and moderate Reynolds numbers, how do these findings convert to high Reynolds numbers?

10.2 Simple numerical example for a VIV aeroelastic energy harvester

The following is an equation of motion that typically describes the VIV of an oscillating cylinder in the cross direction Y:

$$m\ddot{y} + c\dot{y} + ky = F, \tag{10.1}$$

were c indicates the damping of structures, the structural mass is represented by m, the spring constant is given by k, and, in the transverse direction, the force of the fluid is given by F. In the regime where there is a synchronization of the body's frequency of oscillation with a periodic vortex wake mode (or in other words, as periodic fluid force), the force approximation along with response is given by

$$F(t) = F_0 \sin(\omega t + \phi), \tag{10.2}$$

$$y(t) = y_0 \sin(\omega t), \tag{10.3}$$

where

$$\omega = 2\pi f, \tag{10.4}$$

and f denotes the frequency of body oscillation. The response frequency and amplitude can be derived easily from the above equations. For a chosen dimensionless set of parameters, we formulate the equations as follows:

$$A^* = \frac{1}{4\pi^3} \left(\frac{C_Y \sin\phi}{(m^* + C_A)\sigma} \right) \left(\frac{U^*}{f^*} \right)^2 f^*, \tag{10.5}$$

$$f^* = \sqrt{\frac{m^* + C_A}{m^* + C_{EA}}}, \tag{10.6}$$

where the potential added mass coefficient is represented by C_A, and C_{EA} is the effective added mass which includes the effect due to the total transverse fluid force in-phase with acceleration of the body given by $C_Y \cos\phi$:

$$C_{EA} = \frac{1}{2\pi^3} \frac{C_Y \cos\phi}{A^*} (\frac{U^*}{f^*})^2. \tag{10.7}$$

10.2.1 Existence of a "critical mass"

Many studies show that as the systematic mass decreases, the U^* velocity regime increases with vibrations of large-amplitude. Anthony Leonard suggested a wide scale of those regimes to quite a low mass ratio, depending on computational simulation data, at the ONR meeting at Brown University. Surprisingly, the latest findings demonstrate that the sync scheme is infinitely large, not just if the mass is null, but also if the mass drops below a critical point, the numerical value of which is dependent on the vibrating body form.

The top end of the free vibration sync regime of the cylinder, along with the damping of low mass, is usually characterized by a branch of lower amplitude which consists of surprisingly stable frequency of vibration, f^*_{lower}, and as the mass is reduced the level of the frequency increases. Govardhan et al. [23] presented for the lower frequency branch f^*_{lower} a large data set, which is plotted versus m^*; it yielded the data collapse onto a single curve fit:

$$f^*_{\text{lower}} = \sqrt{\frac{m^* + 1}{m^* - 0.54}}. \tag{10.8}$$

This expression offers a realistic and easy way to measure the highest frequency retrievable via the VIV device in the sync regime when the mass ratio m^* is known. The frequency of vibration is exponentially high with the mass ratio being limited to a value of 0.54. Govardhan et al. [23] thus concluded the existence of a critical mass ratio m^*_{crit},

$$m^*_{\text{crit}} = 0.54002, \tag{10.9}$$

below which for finite velocities it is never possible to achieve the response of the lower branch, U^*. For finite $\frac{U^*}{f^*}$, these terms and conditions shall apply, when the structural mass falls below a critical value. It can be predicted that there exist large amplitude vibrations for the velocities U^* which extend to infinity:

$$U^* \text{ end of synchronization} = 9.25\sqrt{\frac{m^* + 1}{m^* - 0.54}}. \tag{10.10}$$

10.3 Vortex-induced vibrations in circular cylinders

As far as harvesting energy from VIV is concerned, a recognized definition is at the piezoelectric cantilever beam's end where a circular cylinder is connected. At the point of low Reynolds numbers (depending upon the circular cylinder's diameter), the flow becomes precisely symmetrical, as the potential theory predicted. Moreover, with the increase in Reynolds number, the flow becomes asymmetrical, and therefore a phenomenon known as Karman vortex street happens. These vortices are shed alternatively, and the body has unsteady aerodynamic forces. Once these vortices are drained at a frequency greater than the harvester's normal frequency, they are locked or synchronized and transverse resonant vibrations occur [24–28]. They are called vortex-induced vibrations (VIVs). Fig. 10.1 indicates the variance of the amplitude of oscillation relative to the variations in the flow rate and the existence of resonant shedding frequency corresponding to the cylinder's natural frequency. Williamson and his co-authors [29–32] have shown that different modes of response can take place based on the mass-damping parameter.

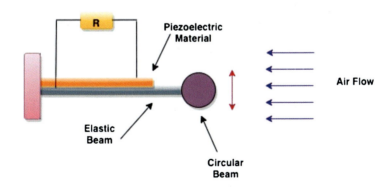

Figure 10.1 Schematic of a wake VIV-based piezoelectric energy harvester.

10.4 Energy harvesting form the PZT based on VIV

The equations governing the beam cantilever supporting a bluff body according to Euler–Bernoulli beam theory can be written as:

$$\rho_x \frac{\partial \omega^2(x, t)}{\partial t^2} + C_\alpha \frac{\partial \omega(x, t)}{\partial t} +$$
$$C_s I \frac{\partial \omega^5(x, t)}{\partial x^4 \partial t} + EI \frac{\partial \omega^4(x, t)}{\partial x^4}$$
$$+ \Theta_P \left(\frac{d\delta(x - I_{p1})}{dx} - \frac{d\delta(x - I_{p2})}{dx} \right) V_{PZT} = f(x, t), \tag{10.11}$$

where the deflection of the beam with respect to x and y coordinates is denoted by $\omega(x, t)$, the beam's per unit length mass is given by ρ_x, viscous damping coefficient is C_a, and the strain rate damping coefficient is C_s. The moment of inertia cross-section is defined by I, whereas EI is PZT beam's bending stiffness which includes contributions from both the beam and PZT. The PZT coupling term is given by Θ_P. At point x, the Dirac delta function is given by $\delta(x)$; the start and end points of PZT on the beam are given as L_{P1} and L_{P2}. Related to the aerodynamics force of the bluff body and its inertia, the external force which is imposed on the beam is denoted by $f(x, t)$,

$$f(x, t) = F_{\text{tip}} \delta(x - L_b) - [0.5DF_{\text{tip}} - I_b \frac{\partial w^3(L_b, t)}{\partial x \partial t^2}] \frac{d\delta(x - L_b)}{dx} \tag{10.12}$$

and

$$F_{\text{tip}} = F_\alpha + F_{\text{impact}} - M_b \frac{\partial \omega^2(L_b, t)}{\partial t^2} - 0.5DM_b \frac{\partial \omega^3(L_b, t)}{\partial x \partial t^2}, \tag{10.13}$$

where I_b and M_b are the rotational moments and center of mass of bluff body, respectively; F_α is the aerodynamic force, resulting from the vortex shedding as the airflow travels across the cylinder's bluff body. The force of impact from the ball is represented by F_{impact}, when it impacts the membrane and the force is applied on the bluff body.

The modal frequency is usually much higher for a thin cantilever beam than for a higher-order mode's first modal frequency. As this study focuses on the low-frequency band, in which only the first beam vibration mode is dominant, the higher beam vibration mode is not taken into account. Assume that $\omega(x, t) = \phi(x)\eta(t)$, where the shape function is $\phi(x)$ and the first mode's modal coordinate is $\eta(t)$. Suppose y_M is the vibration displacement bluff body along the y-axis,

$$y_M(t) = \omega(L_b, t) + 0.5D \frac{\partial \omega(L_b, t)}{\partial x} = (\phi(L_b) + 0.5D \frac{d\phi(L_b)}{dx})\eta(t). \tag{10.14}$$

It should be noted that the amplitude of $\phi(x)$ is undermined. It is assumed for simplicity that $\phi(x)$ satisfies

$$\phi(L_b) + 0.5D\frac{d\phi(L_b)}{dx} = 1. \tag{10.15}$$

Consequently,

$$\eta(t) = y_M. \tag{10.16}$$

Substituting $\omega(x, t) = \phi(x)\eta(t) = \phi(x)y_M$ and multiplying it with $\phi(x)$ and integrating over beam's length, we can write Eq. (10.11) as a single-degree-of-freedom oscillator:

$$M\ddot{y}_M + C\dot{y}_M + Ky_M + \Theta V_{PZT} = F_a + F_{\text{impact}}, \tag{10.17}$$

$$M = \int_0^{L_b} \rho_x\phi^2(x)\,dx + M_b + I_b(\frac{d\phi(L_b)}{dx})^2, \tag{10.18}$$

$$C = \int_0^{L_b} [c_a\phi^2(x) + c_s I\phi(x)\frac{d\phi^4(x)}{dx^4}]dx, \tag{10.19}$$

$$K = \int_0^{L_b} EI\phi(x)\frac{d\phi^4(x)}{dx^4}\,dx, \tag{10.20}$$

$$\Theta = \Theta_p[\phi'(L_{p2}) - \phi'(L_{p1})]. \tag{10.21}$$

The free oscillation test and the static deflection test can be used instead of computerizing the modus type function on the basis of the boundary condition of the cantilever beam that supports a tip mass M, equivalent stiffness K, equivalent damping C, and electromechanical coupling coefficient Θ.

In accordance with a wake oscillator model, the aerodynamic force F_a is written when the bluff cylinder is bypassed by the airflow:

$$F_a = \frac{1}{4}\rho_0 DU^2 HC_{L_0}q - \frac{1}{2}\rho_0 UDHC_D\dot{y}_M. \tag{10.22}$$

In the above equation, air density is $\rho_0 = 1.2041$ kg/m^3, the incoming wind speed is indicated by U, the steady mean lift coefficient is given by $C_{L_0} = 0.3$, the steady mean drag coefficient is $C_D = 1.2$, and $q(t)$ is the intermediate VIV aerodynamics represented by the imaginary variable of a wake oscillator model, i.e.,

$$\ddot{q} + \lambda_{\text{shed}}(q^2 - 1)\dot{q} + \omega_{\text{shed}}^2 q = \frac{A}{D}\ddot{y}_M. \tag{10.23}$$

188 Piezoelectric Aeroelastic Energy Harvesting

VIV is depicted by the wake oscillator model having constants A and λ, the vortex shedding frequency is $\omega_{\text{shed}} = 2\pi S_r U/D$, and Strouhal number is $S_r = 0.2$.

The force of impact F_{impact} is given with the ball's impact on the membrane which is less than the aerodynamic force because of the assumption $m \ll M$; decoupling of aerodynamic force and impact force is done by removing F_{impact}. Thus,

$$M\ddot{y}_M + (C + \frac{1}{2}\rho_0 UDHC_D)\dot{y}_M + Ky_M + \Theta V_{PZT} = \frac{1}{4}\rho_0 DU^2 HC_{L_0} q. \quad (10.24)$$

When a load resistor R_L shunts the PZT, the electrical equation of PZT is given as

$$\frac{V_{PZT}}{R_L} + C_P \dot{V}_{PZT} - \Theta \dot{y}_M = 0. \quad (10.25)$$

Clamped capacitance is denoted by C_p. Parameters A and λ are identified by matching wind tunnel simulation and experimental data.

Define

$$\gamma^* = \frac{y_M}{D}, \quad (10.26)$$

$$\zeta = \frac{C}{2\sqrt{MK}}, \quad (10.27)$$

$$\zeta_a = \frac{\rho_0 DHC_D}{4\sqrt{MK}}, \quad (10.28)$$

$$\Omega = \sqrt{\frac{K}{M}}, \quad (10.29)$$

$$\theta = \frac{\Theta}{MD}, \quad (10.30)$$

$$\sigma = \frac{\rho_0 HC_{L_0}}{4M}, \quad (10.31)$$

$$\xi_1 = \frac{1}{R_L D}, \quad (10.32)$$

$$\xi_2 = \frac{C_p}{D}, \quad (10.33)$$

$$\ddot{\gamma}^* + 2(\zeta + \zeta_a U)\omega\dot{\gamma}^* + \omega^2\gamma^* + \theta V_{PZT} = \sigma U^2 q, \quad (10.34)$$

$$\ddot{q} + \lambda\omega_{\text{shed}}^2 q = A\ddot{\gamma}^*, \quad (10.35)$$

$$\zeta_1 V_{PZT} + \zeta_2 \dot{V}_{PZT} - \Theta\dot{\gamma}^* = 0. \quad (10.36)$$

The average output power is

$$P_{PZT} = \frac{\int_{t_1}^{t_2} V_{PZT}^2(t)\,dt}{(t_2 - t_1)R_L},$$ (10.37)

where the simulation's start and end times are given by t_1 and t_2.

10.5 VIV-based energy harvesters

In this section a overview of different kind of VIV-based energy harvester is provided. Piezoelectric-based aeroelastic energy harvesting is of great interest for many researchers [33–39]. In 2001 Aellen et al. [40] first discussed the consequences of vortex-induced vibrations on the membrane which is flexible with PVDF [41–44]. The authors stated the capability of membranes to display lock-in behavior, when the shedding frequency becomes the same with the basic bluff body frequency. There exist no reports of electrical energy extracted and harvester's efficiency in their work. The "eel" concept was implemented for energy harvesting from the waves of the ocean by Taylor et al. [45]. The prototype "eel" was developed, and in a flow tank it was tested with dimensions: 9.5 inches long, 3 inches wide, and 150 μm thick. An improvement of harvested power was demonstrated when the flapping frequency value reached the vortex shedding frequency.

In their analysis, no distinction was made between simulations and experimental effects of displacement, production of voltage, and harvested power output. Just like "eel" created by Pobering et al. [46], Taylor et al. [45] introduced the capacity of the harvesting energy from the PVDF flag from moving water.

In 2009, Sanchez-Sanz et al. [47] examined the potential for energy harvesting with forces produced by the Karman vortex street around the rectangular micro-prism in the laminar flow regime. They provided recommendations for the construction of devices for harvesting energy without providing any information on the production process.

A PVDF film was mounted to the harvester with a flexible diaphragm. Water flew in the chamber which was below this diaphragm. Inside the chamber, pressure allowed the upward deflection of the diaphragm. The motion of oscillation of the diaphragm with the PZT film thus made it possible to produce electricity. It was stated that the highest harvested voltage was approximately 2.2 V_{pp} and the instantaneous power possible to

190 Piezoelectric Aeroelastic Energy Harvesting

harvest was approximately 0.2 μW. The pressure of excitation oscillated at a 1.196 kPa amplitude and at a frequency of approximately 26 Hz. In a related analysis, Wang used a nearly identical technique to add a bluff body to the chamber's middle. They demonstrated the resulting vortex street as water flew past the bluff body. The diaphragm movement bent the piezoelectric film to produce electricity. Their experimental findings showed that the maximal open-circuit voltage was 0.12 V_{pp} and the instantaneous power was approximately 0.7 nW, whereas with 0.3 kPa amplitude the oscillations of the pressure occurred with 52 Hz frequency.

10.5.1 Circular cylinder based VIV energy harvesters

Some scientists have focused on the possibilities for the use of circular cylinders for energy harvesting from the Karman street vortex. Turbulent layers can harvest energy if the boundaries and circular cylinders wake at a large Reynolds number. Considering the piezoelectric transduction, this was investigated by Akaydin et al. [48] in 2010. A theoretical system that integrates electrical, as well as structural, control equations along with a fluid-flow solver has been developed. Both numerical calculations and experimental results demonstrated strong compliance with their expected open-circuit voltages.

10.5.2 Energy harvester with turbulent flow

In response to their prior investigation, Akaydin et al. [49] experimented with the concept of piezoelectric energy harvesting in its time and space scale, from a turbulent, highly coherent flow. They considered a thin, flexible beam made up of a PVDF coating and a Mylar substrate to have several orientations.

Experimentally it was shown that with the tuning of the frequency of flow shedding along with the piezoelectric generator's natural frequency, it is possible to maximize the generated voltage. They also stated that a small difference of two frequencies, forming significant decreases in extracted power value and, in particular, with shedding frequency lower than harvester's normal frequency.

Both mechanisms, if combined with the beam's resonating conditions, can result in the enhanced energy harvesting, as indicated by the authors. It was suggested that the beam cylinder distance had a substantial effect on the harvester's response. Moreover, the calculation revealed that when the beam cylinder distance is set to 2D, the maximum power harvested was approximately 4 µW. They clarified that near the cylinder maximum, power harvesting is not possible because at these locations there is no full formation of large vortical structures.

10.5.3 Usage of poled and electrode flexible ceramic cylinder for power harvesting

Harvesting power from a poled and electrode flexible ceramic cylinder was theoretically investigated by Xie et al. [50] in 2011. They also developed a 1D mathematical model for the determination of efficiency and power harvested by the aeroelastic harvester. It was said, with shedding frequency coming close to the cylinder's minimum flexural resonant frequency, that the power harvested lies in a range of 1 mW for a cylinder 40 cm in length and 1 cm in diameter in the air with 5 m/s flow speed.

10.5.4 Energy harvesting from a rigid, elastically mounted spherical cylinder

Barrero-Gil et al. [51] logically studied the idea of energy harvesting from a rigid, elastically mounted spherical cylinder. Without including any transduction function, a mechanical model with a lumped-parameter was taken into consideration. The effects of the mechanical damping ratio (ζ) and mass ratio (m^*) on a conversion factor of energy of the aeroelastic system were investigated by the authors. Their study found that the mass-damping parameter ($m^*\zeta$) can have an optimum value so that it maximizes the energy conversion factor. According to them, the most influential parameter is a mass ratio which affects the reduced velocities' region with maximum efficiency.

10.5.5 Energy harvester composed of aluminum cantilevered shim

Akaydin et al. [52] presented a piezoaeroelastic energy harvester composed of an aluminum cantilevered shim covered in PZT material near the base and with a circular cylinder on its tip. The maximum harvested power value was stated to be 0.1 mW at a wind speed of around 1.192 m/s. It should be noted that no computational or analytical modeling was carried out in this analysis to verify the experimental measurements obtained.

10.5.6 Energy harvester based on electromagnetic induction

Wang et al. fabricated and tested a VIV-based aeroelastic energy harvester using electromagnetic induction. They placed the harvester in a water flow channel behind a trapezoidal bluff body. An empirical method was developed to measure the impact of pressure loads in order to test the efficiency of the harvester, device dimensions, and material properties. A prototype of 39 cm^3 volume was considered. Experiments showed that it is possible to generate the output power of order 1.77 µW when excitation pressure oscillates with 0.3 kPa amplitude and 62 Hz frequency.

10.5.7 Vortex-induced vibration aquatic clean energy

Bernitsas et al. [53] patented one of the most important works on energy harvesting from the VIV phenomenon. They named it vortex-induced vibration aquatic clean energy (VIVACE). This makes use of a passive circular cylinder with upwards and downwards motions caused by the vortex shedding. They mounted this passive cylinder on the springs, and electricity was generated in a range of velocities of the flow. Mainly, there is a phenomenon of conversion into electricity from mechanical energy via linear generators or rotary equipment.

10.5.8 Passive turbulence control

To improve harnessed power of the circular cylinder based on VIV, a creative method was presented by MRELAB [54], called passive turbulence control (PTC). For the circular cylinder, PTC is the roughness selectively distributed over the surface which provides transverse galloping after the upper branch of VIV. The range of oscillation amplitude increases up to 3 diameters in the transverse galloping whereas the value of this range is 1.65D if it is in synchronization [54]. At the time of transverse galloping, oscillation frequency decreases to some extent. The amplitude is directly related to the efficiency, whereas if f^* is reduced it can cause a decrease in the efficiency.

10.5.9 Harvester's performance enhancement

Weinstein et al. [55] added a fin on the piezoelectric bender's tip and studied the influence for the increment of performance of the harvester based on VIV. Excitation of the harvester was considered in a heating, ventilation, and air conditioning (HVAC) flow. It has been reported that the wind speeds from 2 to 5 m/s are resonant with little weight along with the results of a fin. The authors claimed that their suggested harvester has the ability to be successfully put in and around HVAC systems and then it works as a node's sensor and wireless radio.

Fluctuating lift force representation was determined by some researchers [56–59] who solved Navier–Stokes equations with the help of direct numerical simulations (DNS). In such numerical simulations, the cylinder's motion and induced voltage cannot be resolved without the determination of fluid loads.

Yet, without understanding the cylinder's motion and the influence of harvested energy from this motion, the fluid load's calculation is not possible. There is a dire need because of these complications for the fluid force equation governing the flow field dynamics, the response equations of the cylinder, and voltage equation controlling the cylinder dynamics; and the voltage generated must be dealt with in a coupled manner.

Hamming's fourth-order predictor–corrector was used to overcome this difficulty [57,60]. This technique couples the fluid loads with ODEs which govern the voltage and cylinder's motion. The state of prediction of the cylinder is dependent on the load resistance, and the new fluid loads are computed with CFD code. For more on this technique, see Mehmood et al. [61].

Increasing the precision of the VIV phenomena models becomes costly in terms of time and computational power. It is highly complicated to undergo such simulations if 3D effects, elasticity, and turbulence mechanisms are taken into consideration. As a result, they ares assumed to make clear and reliable the predictions of VIV phenomena and the response of the structure. The commonly used solution, known as the wake oscillator model, was first introduced by Birkhoff et al. [62].

Hartlen et al. [63] and Skop et al. [28] have modeled the lift using a nonlinear oscillator equation. Subsequently, Skop et al. [64] contributed to the extraordinary observations about the wave oscillator model's workings.

In the definition of the cross-flow force, Facchinetti et al. [65] recommended a stall term, while Facchinetti et al. [65] extensively studied the force distribution caused by the motion of the cylinder on the wake oscillator. Facchinetti et al. [65] have discovered that the acceleration of the coupling representation is best for modeling the lock-in area. This was verified by Violette et al. [66] by comparing with experiments and statistical simulation.

10.5.10 Harvesting energy from rigid circular cylinder

Using the principle of VIV energy harvesting with a rigid circular cylinder, Abdelkefi et al. [28] modeled the lift of the oscillating cylinder, and the Gauss law was used to design a relation between the output voltage and motion of the cylinder.

On the basis of linear analysis, the load resistance was shown to affect the onset of the synchronization area and its characteristics, as determined by Akaydin et al. [52]. They claimed that with maximization of the electrome-chanical damping there exists a particular load resistance. Furthermore, they demonstrated that increasing the load resistance at higher free stream veloc-ities shifts the onset of synchronization. Theoretical studies have shown that there is a hardening behavior as a result of a non-linearity associated with vortex-induced oscillations with hysteresis regions included.

The study shows that harnessed power is of the order of 1 mW supplied to a load resistance of 1 MΩ. These are the results of the numerical simu-lations. VIV-based energy harvesting from long tensioned cables was taken into consideration by [67–69]. For modeling the fluctuating aerodynamic force, the authors used a wake-oscillator model and, for the mechanical part, they used a lumped parameter model. First, they studied the case of a rigid, elastically mounted cylinder. Afterwards an infinite cable was consid-ered.

When the natural and shedding frequencies match, the power harvested from the VIV-based energy harvester reaches the maximum levels. Their results also found that for all configurations considered, the maximum ef-ficiency is of the same order. Harvesting energy from the VIV circular cylinder was studied by Mehmood et al. [61], when with respect to the transverse displacement degree of freedom a piezoelectric material is at-tached. TA solution for the Navier–Stokes equations was identified for the fluctuating lift force, with the fourth-order Hamming predictor–corrector used to compute the system reaction.

Based on a linear analysis, it has been shown that an increase in elec-trical load resistance is followed by an increase in the coupled frequency of the harvester. The damping was shown to be very low at the lowest load resistance values, to a limit of about 200 k for load resistance, and decreased and stayed low as the load resistance was increased.

The oscillations' amplitude for both the lift coefficient and cylinder's displacement is high for the small values of load resistance, it decreases for a number of load resistances and again increases at higher load resistances. The authors also stated that with the increase in load resistance, the voltage output continuously increases. However, for $R \geq 500$ kΩ there is a smaller increase rate. These findings also revealed that the electrical resistance has an optimum value at which the harvested power for various Reynolds numbers can be maximized.

Mackowski et al. [70] have studied the impact of non-linear restoration forces on VIV energy harvester's response. The harvester comprised a VIV cylinder with a non-linear and linear structural power backed by a water channel. There can be a substantial power increase with a suitable option of non-linear structural restoring forces for wind speeds in wider range compared to its linear counterpart.

Note that in all previously mentioned harvester's which were based on vortex-induced vibrators, only lumped parameter models were used. Based on VIV energy harvester's Euler–Lagrange principle, Dai et al. [71] proposed a non-linear distributed-parameter model. The model presented by Facchinetti et al. [65] was used to present a fluctuating lift coefficient.

It was presented that Galerkin discretization is appropriate to consider at least four modes for the correct prediction of harvester's performance. A very strong agreement was found in the experimental measurements and the derived electroaeroelastic model given by Akaydin et al. [52]. They also stated that at different operating wind speeds, there occurs a compromise between the tip mass of the cylinder and the load resistance which was used to design an efficient energy harvester based on VIV.

The principle of piezoelectric energy harvester was tested using a validated non-linear distributed parameter model to simultaneously harvest energy from vortex-induced vibrations and base excitation [25,72]. The effects on the harvester's performance, of base acceleration, wind speed, and load resistance were evaluated by performing non-linear and linear analysis.

The related electromechanical damping has been shown to improve when the wind speed is in pre- or post-synchronization regions, and therefore the power harvested is reduced. However, when the wind speed is in the field of synchronization, it was recorded that the output power would increase dramatically to 150% compared to two individual harvesters. Increased base acceleration was shown, along with a reduction in vibrational effects caused by the vortex and important effects of the quenching phenomenon by Nayfeh et al. [73].

References

[1] Turgut Sarpkaya, Vortex-induced oscillations: a selective review, 1979.

196 Piezoelectric Aeroelastic Energy Harvesting

[2] O.M. Griffin, S.E. Ramberg, Some recent studies of vortex shedding with application to marine tubulars and risers, 1982.

[3] Peter W. Bearman, Vortex shedding from oscillating bluff bodies, Annual Review of Fluid Mechanics 16 (1984) 195–222.

[4] Geoffrey Parkinson, Phenomena and modelling of flow-induced vibrations of bluff bodies, Progress in Aerospace Sciences 26 (2) (1989) 169–224.

[5] Petros Anagnostopoulos, Flow-Induced Vibrations in Engineering Practice, vol. 31, Wit Pr/Computational Mechanics, 2002.

[6] R.D. Blevins, Flow-Induced Vibrations, Van Nostrand Reinhold, New York, 1990.

[7] E. Naudascher, D. Rockwell, Flow-Induced Vibrations. An Engineering Guide, A.A. Balkema, Rotterdam, Holland, 1994.

[8] Jorgen Fredsoe, B. Mutlu Sumer, Hydrodynamics Around Cylindrical Structures, revised edition, vol. 26, World Scientific, 2006.

[9] Charles HK Williamson, Anatol Roshko, Vortex formation in the wake of an oscillating cylinder, Journal of Fluids and Structures 2 (4) (1988) 355–381.

[10] Mohsin Ali Marwat, Weigang Ma, Pengyuan Fan, Hassan Elahi, Chanatip Samart, Bo Nan, Hua Tan, David Salamon, Baohua Ye, Haibo Zhang, Ultrahigh energy density and thermal stability in sandwich-structured nanocomposites with dopamine@Ag@BaTiO$_3$, Energy Storage Materials 31 (2020) 492–504.

[11] Ahsan Ali, Riffat Asim Pasha, Hassan Elahi, Muhammad Abdullah Sheeraz, Saima Bibi, Zain Ul Hassan, Marco Eugeni, Paolo Gaudenzi, Investigation of deformation in bimorph piezoelectric actuator: analytical, numerical and experimental approach, Integrated Ferroelectrics 201 (1) (2019) 94–109.

[12] Muhammad Usman Khan, Zubair Butt, Hassan Elahi, Waqas Asghar, Zulkarnain Abbas, Muhammad Shoaib, M. Anser Bashir, Deflection of coupled elasticity–electrostatic bimorph PVDF material: theoretical, FEM and experimental verification, Microsystem Technologies 25 (8) (2019) 3235–3242.

[13] Vittorio Memmolo, Hassan Elahi, Marco Eugeni, Ernesto Monaco, Fabrizio Ricci, Michele Pasquali, Paolo Gaudenzi, Experimental and numerical investigation of PZT response in composite structures with variable degradation levels, Journal of Materials Engineering and Performance 28 (6) (2019) 3239–3246.

[14] H. Elahi, A. Israr, R.F. Swati, H.M. Khan, A. Tamoor, Stability of piezoelectric material for suspension applications, in: 2017 Fifth International Conference on Aerospace Science & Engineering (ICASE), IEEE, 2017, pp. 1–5.

[15] Hassan Elahi, Marco Eugeni, Paolo Gaudenzi, Electromechanical degradation of piezoelectric patches, in: Analysis and modelling of advanced structures and smart systems, Springer, 2018, pp. 35–44.

[16] Owen M. Griffin, Steven E. Ramberg, Vortex shedding from a cylinder vibrating in line with an incident uniform flow, Journal of Fluid Mechanics 75 (2) (1976) 257–271.

[17] A. Ongoren, D. Rockwell, Flow structure from an oscillating cylinder. Part 2. Mode competition in the near wake, Journal of Fluid Mechanics 191 (1988) 225–245.

[18] R. Zdero, Ö.F. Turan, D.G. Havard, Toward understanding galloping: near-wake study of oscillating smooth and stranded circular cylinders in forced motion, Experimental Thermal and Fluid Science 10 (1) (1995) 28–43.

[19] Hassan Elahi, Khushboo Munir, Marco Eugeni, Muneeb Abrar, Asif Khan, Adeel Arshad, Paolo Gaudenzi, A review on applications of piezoelectric materials in aerospace industry, Integrated Ferroelectrics 211 (1) (2020) 25–44.

[20] Hassan Elahi, Marco Eugeni, Paolo Gaudenzi, Madiha Gul, Raees Fida Swati, Piezoelectric thermo electromechanical energy harvester for reconnaissance satellite structure, Microsystem Technologies 25 (2) (2019) 665–672.

[21] Hassan Elahi, Marco Eugeni, Paolo Gaudenzi, Faisal Qayyum, Raees Fida Swati, Hayat Muhammad Khan, Response of piezoelectric materials on thermomechanical shocking and electrical shocking for aerospace applications, Microsystem Technologies 24 (9) (2018) 3791–3798.

[22] Hassan Elahi, Zubair Butt, Marco Eugnei, Paolo Gaudenzi, Asif Israr, Effects of variable resistance on smart structures of cubic reconnaissance satellites in various thermal and frequency shocking conditions, Journal of Mechanical Science and Technology 31 (9) (2017) 4151–4157.

[23] R. Govardhan, C.H.K. Williamson, Modes of vortex formation and frequency response of a freely vibrating cylinder, Journal of Fluid Mechanics 420 (2000) 85–130.

[24] A. Baz, J. Ro, Active control of flow-induced vibrations of a flexible cylinder using direct velocity feedback, Journal of Sound and Vibration 146 (1) (1991) 33–45.

[25] H.L. Dai, A. Abdelkefi, L. Wang, Modeling and nonlinear dynamics of fluid-conveying risers under hybrid excitations, International Journal of Engineering Science 81 (2014) 1–14.

[26] Viung-Chung Mei, I.G. Currie, Flow separation on a vibrating circular cylinder, The Physics of Fluids 12 (11) (1969) 2248–2254.

[27] G.V. Parkinson, Mathematical models of flow-induced vibrations of bluff bodies, in: Flow-Induced Structural Vibrations, Springer-Verlag, Berlin, 1974, pp. 81–127.

[28] R.A. Skop, O.M. Griffin, On a theory for the vortex-excited oscillations of flexible cylindrical structures, Journal of Sound and Vibration 41 (3) (1975) 263–274.

[29] R. Govardhan, C.H.K. Williamson, Critical mass in vortex-induced vibration of a cylinder, European Journal of Mechanics-B/Fluids 23 (1) (2004) 17–27.

[30] Asif Khalak, Charles HK Williamson, Investigation of relative effects of mass and damping in vortex-induced vibration of a circular cylinder, Journal of Wind Engineering and Industrial Aerodynamics 69 (1997) 341–350.

[31] Asif Khalak, Charles HK Williamson, Motions, forces and mode transitions in vortex-induced vibrations at low mass-damping, Journal of Fluids and Structures 13 (7–8) (1999) 813–851.

[32] C.H.K. Williamson, R. Govardhan, Vortex-induced vibrations, Annual Review of Fluid Mechanics 36 (2004) 413–455.

[33] Hassan Elahi, Khushboo Munir, Marco Eugeni, Paolo Gaudenzi, Reliability risk analysis for the aeroelastic piezoelectric energy harvesters, Integrated Ferroelectrics 212 (1) (2020) 156–169.

[34] Hassan Elahi, Marco Eugeni, Luca Lampani, Paolo Gaudenzi, Modeling and design of a piezoelectric nonlinear aeroelastic energy harvester, Integrated Ferroelectrics 211 (1) (2020) 132–151.

[35] Hassan Elahi, Marco Eugeni, Federico Fune, Luca Lampani, Franco Mastroddi, Giovanni Paolo Romano, Paolo Gaudenzi, Performance evaluation of a piezoelectric energy harvester based on flag-flutter, Micromachines 11 (10) (2020) 933.

[36] Hassan Elahi, The investigation on structural health monitoring of aerospace structures via piezoelectric aeroelastic energy harvesting, Microsystem Technologies (2020) 1–9.

[37] Marco Eugeni, Hassan Elahi, Federico Fune, Luca Lampani, Franco Mastroddi, Giovanni Paolo Romano, Paolo Gaudenzi, Numerical and experimental investigation of piezoelectric energy harvester based on flag-flutter, Aerospace Science and Technology 97 (2020) 105634.

[38] Hassan Elahi, Marco Eugeni, Paolo Gaudenzi, Design and performance evaluation of a piezoelectric aeroelastic energy harvester based on the limit cycle oscillation phenomenon, Acta Astronautica 157 (2019) 233–240.

[39] Marco Eugeni, Hassan Elahi, Federico Fune, Luca Lampani, Franco Mastroddi, Giovanni Paolo Romano, Paolo Gaudenzi, Experimental evaluation of piezoelectric energy harvester based on flag-flutter, in: Conference of the Italian Association of Theoretical and Applied Mechanics, Springer, 2019, pp. 807–816.

[40] J.J. Allen, A.J. Smits, Energy harvesting eel, Journal of Fluids and Structures 15 (3) (2001) 629–640.

[41] Hassan Elahi, M. Rizwan Mughal, Marco Eugeni, Faisal Qayyum, Asif Israr, Ahsan Ali, Khushboo Munir, Jaan Praks, Paolo Gaudenzi, Characterization and implementation of a piezoelectric energy harvester configuration: analytical, numerical and experimental approach, Integrated Ferroelectrics 212 (1) (2020) 39–60.

[42] Hassan Elahi, Khushboo Munir, Marco Eugeni, Sofiane Atek, Paolo Gaudenzi, Energy harvesting towards self-powered IoT devices, Energies 13 (21) (2020) 5528.

[43] Hassan Elahi, Marco Eugeni, Paolo Gaudenzi, A review on mechanisms for piezoelectric-based energy harvesters, Energies 11 (7) (2018) 1850.

[44] Zubair Butt, Riffat Asim Pasha, Faisal Qayyum, Zeeshan Anjum, Nasir Ahmad, Hassan Elahi, Generation of electrical energy using lead zirconate titanate (PZT-5a) piezoelectric material: analytical, numerical and experimental verifications, Journal of Mechanical Science and Technology 30 (8) (2016) 3553–3558.

[45] George W. Taylor, Joseph R. Burns, S.A. Kammann, William B. Powers, Thomas R. Welsh, The energy harvesting eel: a small subsurface ocean/river power generator, IEEE Journal of Oceanic Engineering 26 (4) (2001) 539–547.

[46] Sebastian Pobering, Norbert Schwesinger, A novel hydropower harvesting device, in: 2004 International Conference on MEMS, NANO and Smart Systems (ICMENS'04), IEEE, 2004, pp. 480–485.

[47] Mario Sanchez-Sanz, Belen Fernandez, Angel Velazquez, Energy-harvesting microresonator based on the forces generated by the Kármán street around a rectangular prism, Journal of Microelectromechanical Systems 18 (2) (2009) 449–457.

[48] H.D. Akaydın, Niell Elvin, Yiannis Andreopoulos, Wake of a cylinder: a paradigm for energy harvesting with piezoelectric materials, Experiments in Fluids 49 (1) (2010) 291–304.

[49] Huseyin Dogus Akaydin, Niell Elvin, Yiannis Andreopoulos, Energy harvesting from highly unsteady fluid flows using piezoelectric materials, Journal of Intelligent Material Systems and Structures 21 (13) (2010) 1263–1278.

[50] Jiemin Xie, Jiashi Yang, Hongping Hu, Yuantai Hu, Xuedong Chen, A piezoelectric energy harvester based on flow-induced flexural vibration of a circular cylinder, Journal of Intelligent Material Systems and Structures 23 (2) (2012) 135–139.

[51] Antonio Barrero-Gil, Santiago Pindado, Sergio Avila, Extracting energy from vortex-induced vibrations: a parametric study, Applied Mathematical Modelling 36 (7) (2012) 3153–3160.

[52] H.D. Akaydın, N. Elvin, Y. Andreopoulos, The performance of a self-excited fluidic energy harvester, Smart Materials and Structures 21 (2) (2012) 025007.

[53] Michael M. Bernitsas, Kamaldev Raghavan, Y. Ben-Simon, E.M.H. Garcia, Vivace (vortex induced vibration aquatic clean energy): a new concept in generation of clean and renewable energy from fluid flow, Journal of Offshore Mechanics and Arctic Engineering 130 (4) (2008).

[54] Che-Chun Jim Chang, R. Ajith Kumar, Michael M. Bernitsas, VIV and galloping of single circular cylinder with surface roughness at $3.0 \times 10^4 \leqslant Re \leqslant 1.2 \times 10^5$, Ocean Engineering 38 (16) (2011) 1713–1732.

[55] Lee A. Weinstein, Martin R. Cacan, P.M. So, P.K. Wright, Vortex shedding induced energy harvesting from piezoelectric materials in heating, ventilation and air conditioning flows, Smart Materials and Structures 21 (4) (2012) 045003.

[56] Imran Akhtar, Parallel simulations, reduced-order modeling, and feedback control of vortex shedding using fluidic actuators, PhD thesis, Virginia Tech, 2008.

[57] A. Mehmood, A. Abdelkefi, I. Akhtar, A.H. Nayfeh, A. Nuhait, M.R. Hajj, Linear and nonlinear active feedback controls for vortex-induced vibrations of circular cylinders, Journal of Vibration and Control 20 (8) (2014) 1137–1147.

[58] Karl W. Schulz, Yannis Kallinderis, Unsteady flow structure interaction for incompressible flows using deformable hybrid grids, Journal of Computational Physics 143 (2) (1998) 569–597.

[59] J. Yang, S. Preidikman, E. Balaras, A strongly coupled, embedded-boundary method for fluid–structure interactions of elastically mounted rigid bodies, Journal of Fluids and Structures 24 (2) (2008) 167–182.

[60] Brice Carnahan, Herbert A. Luther, Applied numerical methods, Technical report, 1969.

[61] A. Mehmood, A. Abdelkefi, M.R. Hajj, A.H. Nayfeh, I. Akhtar, A.O. Nuhait, Piezoelectric energy harvesting from vortex-induced vibrations of circular cylinder, Journal of Sound and Vibration 332 (19) (2013) 4656–4667.

[62] Garrett Birkhoff, E.H. Zarantonello, Wakes, Jets and Cavities, New York, 1957.

[63] Ronald T. Hartlen, Iain G. Currie, Lift-oscillator model of vortex-induced vibration, Journal of the Engineering Mechanics Division 96 (5) (1970) 577–591.

[64] R.A. Skop, S. Balasubramanian, A new twist on an old model for vortex-excited vibrations, Journal of Fluids and Structures 11 (4) (1997) 395–412.

[65] Matteo Luca Facchinetti, Emmanuel De Langre, Francis Biolley, Coupling of structure and wake oscillators in vortex-induced vibrations, Journal of Fluids and Structures 19 (2) (2004) 123–140.

[66] R. Violette, Emmanuel De Langre, J. Szydlowski, Computation of vortex-induced vibrations of long structures using a wake oscillator model: comparison with DNS and experiments, Computers & Structures 85 (11–14) (2007) 1134–1141.

[67] Clément Grouthier, Sébastien Michelin, Rémi Bourguet, Yahya Modarres-Sadeghi, Emmanuel De Langre, On the efficiency of energy harvesting using vortex-induced vibrations of cables, Journal of Fluids and Structures 49 (2014) 427–440.

[68] Clément Grouthier, Sébastien Michelin, Emmanuel de Langre, Optimal energy harvesting by vortex-induced vibrations in cables, in: Proceedings of the 10th FIV 2012 International Conference on Flow-Induced Vibrations Conference (& Flow-Induced Noise), 2012.

[69] Clement Grouthier, Sebastien Michelin, Emmanuel de Langre, Energy harvesting by vortex-induced vibrations in slender structures, in: International Conference on Offshore Mechanics and Arctic Engineering, vol. 55416, American Society of Mechanical Engineers, 2013, V007T08A013.

[70] A.W. Mackowski, C.H.K. Williamson, An experimental investigation of vortex-induced vibration with nonlinear restoring forces, Physics of Fluids 25 (8) (2013) 087101.

[71] H.L. Dai, A. Abdelkefi, L. Wang, W.B. Liu, Control of cross-flow-induced vibrations of square cylinders using linear and nonlinear delayed feedbacks, Nonlinear Dynamics 78 (2) (2014) 907–919.

[72] H.L. Dai, A. Abdelkefi, L. Wang, Piezoelectric energy harvesting from concurrent vortex-induced vibrations and base excitations, Nonlinear Dynamics 77 (3) (2014) 967–981.

[73] Ali Hasan Nayfeh, Dean T. Mook, P. Holmes, Nonlinear oscillations, 1980.

CHAPTER 11

Galloping-based aeroelastic energy harvesting

Contents

11.1. Introduction	201
11.2. Transverse galloping	202
11.2.1 Quasi-steady estimation	203
11.3. Mathematical model of transverse galloping	208
11.4. Wake galloping	209
11.4.1 Types of wake galloping	211
11.4.1.1 Wake galloping with unstable slope	*211*
11.4.1.2 Wake galloping with stable slope	*211*
11.4.2 Turbulence effects	211
11.4.3 Galloping response	212
11.5. Conversion factor	212
11.6. Evaluation of critical conditions with refined and multi-model approaches	213
11.7. Semi-analytical versus numerical solutions: non-linear galloping	215
11.8. Harnessable energy	216
References	216

11.1 Introduction

The idea of using wind energy for energy extraction is not new. However, the interest in energy extraction from the oscillation phenomena caused by the flow is recent [1–7]. An example of this is the energy-harvesting eel. There is a piezoelectric membrane positioned after a bluff. When the membrane's elastic properties and mass are appropriate, significant oscillations are induced in the membrane by the vortex path formed behind the body, these oscillations can then be converted to electricity.

To use galloping vibrations for energy harvesting, a square, D-section, or triangle-shaped prismatic structure is connected to the free end of the piezoelectric cantilever beam. This method is referred to as transverse galloping, known from large amplitude oscillations. Galloping happens when the derivative of the aerodynamic lift's steady-state coefficient becomes negative. This is considered as an essential galloping condition [8]. The bifurcation diagram of varying flow speed and flutters in the airfoil section is the same. The distinction is that a one-degree-of-freedom system experi-

Piezoelectric Aeroelastic Energy Harvesting
https://doi.org/10.1016/B978-0-12-823968-1.00022-2

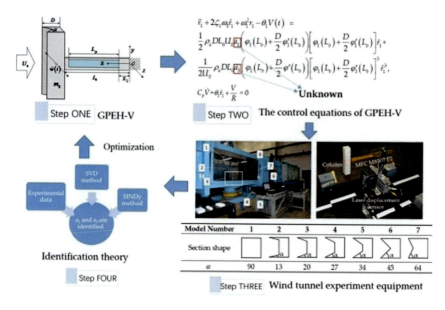

Figure 11.1 Overall mechanism of a galloping-based piezoelectric energy harvester [9].

ences the phenomenon of transverse galloping and a two- or three-degree system is used for airfoil flutter. The overall mechanism for galloping-based piezoelectric energy harvesting is represented in Fig. 11.1 [9].

11.2 Transverse galloping

The transverse galloping is another potential aeroelastic condition [8]. This phenomenon was first researched and explained in 1943. To understand the forces of aerodynamics, the author used a quasi-steady theory. Criteria have been created for the galloping phenomenon in the mentioned part, which has better-specified lift and drag coefficients. In general, the galloping of transverse elastic bluffs occurs as the wind rate expands the critical value that triggers instability and oscillates the prismatic form. Galloping phenomena often induce large amplitudes of oscillations. The use of the piezoelectric transducer is useful since there is a relation of generated voltage with an amplitude of oscillation. Many researches were focused on the various parameters effecting the galloping behavior for different bluff body structures [10–17]. In addition, these studies have investigated the effect of geometry changes in the cross-section on structural amplitudes and the use of transverse galloping for energy extraction [18,19]. The system of mechanical behavior is represented by a model of a lumped-parameter. Polynomial representation-based quasi-steady approximation is taken into consideration. In order to harvest energy, different geometries were studied. Specified approaches on how to harvest this energy have not been explored in this report. Schematic of a galloping-based piezoelectric energy harvester is represented in Fig. 11.2.

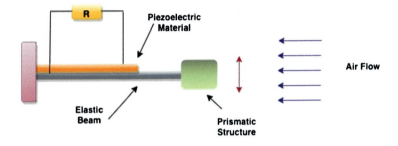

Figure 11.2 Schematic of a galloping-based piezoelectric energy harvester.

11.2.1 Quasi-steady estimation

Near-steady estimation is considered in all studies conducted in galloping-based aeroelastic energy harvesting. The oscillation characteristic time scale in the phenomenon of transverse galloping is $\frac{2\pi}{\omega}$, with ω being the system oscillation frequency, which in general is larger than the flow motion's characteristic time scale of $\frac{D}{U}$, where U represents the incoming flow velocity and D is the body's characteristic dimension, normal to the incoming flow. There is a direct relation of the drag coefficient C_d and lift coefficient C_l with the aerodynamic coefficient C_y, where the angle of attack is denoted by α. The drag and lift coefficients have generally been numerically or experimentally calculated to be functions of the attack angle, and an evaluation of the overall aerodynamic coefficient C_y has been obtained. Multiple researchers have used this technique [20–22]. For applications with a relatively large Reynolds number, an approximation of the aerodynamic force is attainable by using the cubic polynomial function α as

$$F_y = \frac{1}{2}\rho U^2 D(a_1\alpha + a_3\alpha^3), \tag{11.1}$$

where empirical coefficients a_1 and a_3 are obtained by fitting C_y versus α.

Theoretically and experimentally, [23] explored the harvesting of piezoelectric energy from galloping beams with the D-shaped cross-section geometry in 2011. The galloping force and moment were modeled for a quasi-steady approximation. The power output was seen to increase rapidly as the wind speed grew. The authors stated that a wind speed of 5.6 mph was needed to produce energy from this system. The overall harvested potential was also demonstrated to be 1.14 mW at a 10.5 mph wind level. It was claimed that a quasi-steady approximation can be beneficial for the modeling of fluid forces for the galloping phase. It can be seen here that a cubic polynomial was considered to evaluate the harvester's first mode. This expectation leads to an over-estimation or under-estimation of certain parameters, which includes galloping force coefficients with the damping factor.

Following up their last research, the authors [23] proposed the idea of galloping energy harvesting using an equilateral triangle of cantilever beams with piezoelectric sheets linked to the surface. In addition, a linear term combined with cubic polynomial terms was believed to be the first form of the structure. A prototype system was considered which produced more than 50 mW at 11.6 mph wind speed. It has been reported that this amount of harvested power is adequate to supply most commercially available wireless sensors.

In 2012, [24] examined the idea of using square–cylinder galloping for energy harvesting. The main focus was on how the Reynolds number affects the aerodynamic force, the rate of onset of galloping, and the level of harvested power. The device was designed for the coupling of voltage and displacement with a model of the lumped parameter. A quasi-static approximation was employed in modeling the galloping force. Hopf bifurcation's normal form was extracted closer to the galloping process's onset, based on the definition of instability, and for the determination of effects of device parameters on its outputs when it is closer to bifurcation. The findings demonstrated that the resistance to electric loads and the Reynolds number are significant to determine the onset and voltage of the galloping. Furthermore, the highest power extraction levels were reported to be followed by minimal transverse displacements in the case of both configurations of low and high Reynolds numbers.

In 2013, [20] developed a non–linear electroaeroelastic coupled model for piezoelectric galloping. The model has been tested with experimental findings as described by Sirohi et al. [23]. It has also been shown that taking into account the mode shapes of the partly covering cantilever piezoelectric beam and cantilever beam's non–linear contribution, a torsional spring in the frontal region representation on the clamped side is very significant for the creation of an effective analytical model. Furthermore, the findings have shown that the galloping onset speed is greatly influenced by the electrical load resistance. In addition, they discovered that the highest extracted power levels are followed by the minimal values of transverse displacement when changes in the resistance of the electric load are associated with the shunt damping effect. It was also elaborated that the harvester under consideration had a subcritical Hopf bifurcation [21] which at lower wind speeds could be carefully used to produce energy. The authors of [25] performed an experimental study of the effect of cross-section geometry on the efficiency of an energy harvester which follows the principle of galloping. The current coefficients were identified and tested by experimental measurements in the lumped parameter model. For the evaluation of the aerodynamic modes, a quasi-steady hypothesis was used. Experimental findings have shown that the square geometry is superior to the rest of the considered cross-sectional geometries (D-sections, rectangles, triangles, etc.). The achieved power output was of the order of 8.4 mW.

Abdelkefi et al. [26] have researched the effect of the temperature on PZT-5H piezoelectric material properties and on onset galloping speed and level of harvesting capacity of an aeroelastic galloping energy harvester. Galloping energy harvester was composed of a bimorph piezoelectric beam with a tip mass of prismatic structure. To couple the output voltage and displacement, a model with parameters that are non-linear in nature was developed. The temperature variation has been found to have a major effect on the response of the harvester. They suggested that at low wind speed and higher temperature values with various load resistances, energy could be obtained. However, at low temperatures, additional energy can be generated with relatively higher wind speeds.

Zhao et al. [27] submitted a comparative analysis of potential models for the efficiency of piezoaeroelastic energy harvesters focused on galloping. These approaches include the model were based on lumped-parameter, single-mode, and multi-mode Euler–Bernoulli distributed-parameter models. Note that earlier these models were developed and used by many researchers independently [20,25]. A harvester is made up of a piezoelectric cantilever beam connected to a square cross-section tip mass. As described in all the experiments, the galloping force was approximated with the quasi-state approximation. All models considered have been proven to be able to properly predict harvester efficiency. They showed that, because of its simplicity and ease of determining electromechanical coupling conditions, the lumped-parameter model is beneficial. It is important to remember that on the basis of experimental measurements, the lumped model parameters were obtained.

Ali et al. [28] have suggested a galloping-based energy harvester using electromagnetic induction. To a cross-section, they attached a very lightweight box, which was connected to the cantilever beam, known to undergo galloping oscillations with the exposure to the incoming flow rate. In order to test the finite element model, the galloping section lift and drag coefficients were estimated and these aerodynamic coefficients were found to be in line with previous experiments. In terms of achieving the large amplitudes of oscillation and extracted force levels, the authors pointed to the D-section as the strongest cross-section geometry.

A relationship between the wind speed of an energy harvester based on galloping and dimensionless version of harvested power was developed by Bibo et al. [29] in 2014. The construction of total galloping force depended upon quasi-steady hypothesis, and a lumped-parameter model was taken into consideration to link the bluff body's voltage with displacement, along with a polynomial representation. Analytical expressions of the system's output can be approximated by using the method of multiple scales (MMS). With the help of these approximations, the authors stated that it is possible to pose this relationship in a particular curve that depends on the properties of the prismatic structure aerodynamics.

Tang and his colleagues [30,31] used the same method as in the analysis. Zhao et al. [27] suggested a representation of an equivalent circuit method to predict the galloping-based energy harvester's efficiency.

A model of lumped-parameter and a quasi-stationary approach were taken into consideration in order to calculate the response of the harvester. For the galloping force, a user-defined (non-standard) electronic component has been modeled. The proposed model predictions and theoretical calculations were found to be in a very strong agreement, on the one hand, and the experimental measurements, on the other. Zhao et al. [32] suggested a two-degree-of-freedom energy harvester based on galloping with magnetic interaction, cut-out beam, and a square cross-section tip mass. Based on experimental studies, they demonstrated that the starting speed of the galloping could be reduced to 1 m/s in consideration. In addition, it has been shown that the extracted power can be increased for wind speeds below 4.5 m/s, which is exceptional for conditions of HVAC flow and for wireless sensing nodes typically used in indoor monitoring systems. The concept of a bluff body base and galloping (i.e., energy harvesting from hybrid vibrations) was theoretically investigated by Yan et al. [33], whereas Bibo et al. did an experimental investigation along with theoretical analysis [34]. There was a development of electro-aeroelastic models which were non-linear in nature and took the galloping force and moment non-linearities into account, as well as the base excitation effects.

Linear and non-linear analyses were conducted by Yan et al. [35] to assess the effect of the wind speed and electrical load resistance on the coupled frequency, onset velocity of galloping, electromechanical damping, and harvested power level. For the accurate prediction of he harvester's response, if the Galerkin approach is used then there is a need for only one mode shape as demonstrated by Bibo et al. [34]. It has been said that new non-linear processes take place as others tend to vanish, depending upon the connection between base and galloping excitations and values such as load resistance, base acceleration, and wind speed. Following their previous analysis, Yan and Abdelkefi (2014) carried out a non-linear characterization in the case of the pairing of a harvester with aerodynamical and vibratory base excitations. Modern approaches have been used in non-linear dynamics, which include power spectrum, Poincare sections, and phase portraits. It has been reported that different reactions may occur depending upon the relationship of the harvester's excitation frequency with the harvester's global normal frequency. It was demonstrated that the Poincare section and the power spectrum are beneficial for the correct measurement of reactions of the harvester once they are two-periodic, quasi-periodic, and n-periodic. If the phase portrait is used for the characterization of the harvester response then it is possible to obtain erroneous predictions. It was also noted that when the base acceleration is increased, a spectrum of effects of non-galloping is broadened among two pull-out frequencies. The existence of the quenching phenomenon has been used to describe these effects, which leads to either the galloping effect decay or the disappearance when the base excitation is accelerated.

In an investigation of the galloping-based energy harvesting using the phenomenon of electromagnetic induction, Vicente-Ludlam et al. [36] and Dai et al. [37] used a coupled electroaeroelastic model. Vicente-Ludlam et al. [36] based their research on the investigation of the optimum performance for the energy harvesters and stated that for achieving optimal harvested power, electrical load resistance needs to be optimal. The electrical load resistance effects on the harvested power levels and onset galloping speed were investigated by Dai et al. [37].

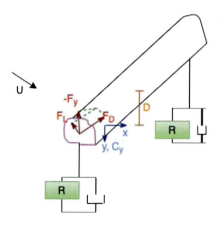

Figure 11.3 Fluid force on the cross-section.

11.3 Mathematical model of transverse galloping

Consider a simple design that consists of a spring-mounted prismatic body that is susceptible to galloping when subjected to the incoming flow in transverse direction; see Fig. 11.3. The design consists of a mechanical damping ratio ζ, m is the mass per unit length, and ω_N is the natural circulation frequency. Furthermore, the body is small enough to assume bidimensional flow, and the event flow is turbulence-free. Then the equation governing the system's dynamics is

$$m(\ddot{y} + 2\zeta\omega_N\dot{y} + \omega_N^2 y) = F_y = \frac{1}{2}\rho U^2 D C_y, \qquad (11.2)$$

where th dot represents differentiation with respect to time, ρ represents the fluid density, and the vertical position is denoted by y. Here the fluid density throughout the analysis will be considered constant. Also D represents the characteristics dimensions of the body normal to flow, incident flow's undisturbed velocity is given by U, while F_y represents the force per unit length of fluid in the direction normal to incident flow, whereas the instantaneous force coefficient of fluid is given by C_y, which is in the transverse direction to incident flow.

Normally, the phenomenon of transverse galloping is characterized by body oscillation timescale given by $\sim \frac{2\pi}{\omega_N}$ which is greater than the characteristic flow timescale $\sim \frac{D}{U}$. Hence,

$$\tan(\alpha) = \frac{\dot{y}}{U}, \qquad (11.3)$$

where $\dot{y}$ is the empirical function of fluid force, which can be approximated by a polynomial when one knows the static variations of C_y with α. It is observed that C_y can be related to the drag and shift coefficients C_d and C_l as

$$C_D C_y = -(C_L + C_D \tan \alpha)/\cos \alpha. \tag{11.4}$$

A cubic polynomial is considered to approximate the force coefficient of vertical fluid,

$$F_y = \frac{1}{2}\rho U^2 D(a_1 \frac{\dot{y}}{U} + a_3 (\frac{\dot{y}}{U})^3), \tag{11.5}$$

where the empirical coefficients represented by a_1 and a_3 are fit by C_y polynomial versus $\tan(\alpha)$ dependence which is measured in the static tests.

The linear coefficient

$$a_1 = -\partial C_L/\partial \alpha + C_D \tag{11.6}$$

is at zero angle of attack to the slope of coefficient vertical fluid force. It is necessary for galloping that $a_1 > 0$ and the lift coefficient slope must be negative. The non-linear dependence of C_y with α is represented by a_3 and is negative. It is noted that there is a limitation when C_y increases with α. There is a dependence of the a_1 and a_3 on several factors such as the cross-section geometry, also on incident flow's characteristics, or $\frac{L}{D}$, the body's aspect ratio.

If dimensionless variables such as $\tau = \omega_N t$ and $\eta = \frac{y}{D}$ are used then, taking the equation for F_y into account, we get

$$\eta'' + 2\zeta\eta' + \eta = \frac{U^{*2}}{2m^*}(a_1 \frac{\eta'}{U^*} + a_3)(\frac{\eta'}{U^*})^3, \tag{11.7}$$

where primes denote differentiation with respect to dimensionless time τ, and the ratio of body's mean density to the surrounding fluid's density, also known as the dimensionless mass ratio, is given by $m^* = \frac{m}{\rho D^2}$, while the reduced velocity is represented by $U^* = \frac{U}{\omega_N D}$.

11.4 Wake galloping

In the event of harnessing energy from wake galloping, a bluff body is attached to the piezoelectric cantilever beam's free end, and another is placed on the front. It has been demonstrated in parallel cylinders that wake galloping phenomena depend on the position of the front cylinders, and vibrations are only present at definite distances between two parallel cylinder systems [38]. In the past few years, an energy harvester from parallel cylinders, known as wake galloping, was considered. Hobbs et al. [39,40] researched piezoelectric production of energy from many elastic cylinders in a line [41]. It has been shown that when the piezoelectric cylinders are clustered together, the currents flow together and the generated flows by cylinders upstream can increase significantly the harvested power level by downstream cylinders. The authors also suggested that a well-defined distance between the cylinders, calculated as $L/D = 3.3$, should increase the extracted power. The authors argued that the wind speed from which the generated power is boosted is an ideal value. The highest value of the generated power for wind speed was associated with $f_s/f_n = 1.6$, this is very unusual because normally the maximum value of harvested power is expected to occur when the shedding frequency (f_s) comes closer to the natural frequency (f_n) of the harvester. Schematic of a wake galloping-based piezoelectric energy harvester is represented in Fig. 11.4. Furthermore,

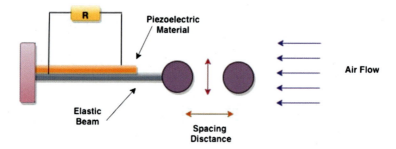

Figure 11.4 Schematic of a wake galloping-based piezoelectric energy harvester.

$$C_y = -[C_l \cos(\alpha) + C_d \sin(\alpha)]. \qquad (11.8)$$

Visually, they tried to clarify why this difference was discovered and mentioned the hypothesis: Oscillating cylinder's vortex shedding frequency is very much like the stationary cylinders and the consequences of the frequency lock-in were still not taken into account; and (b) the vortex shedding occurs in the horizontal plane, which might not be right, considering the comparatively low cylinders' aspect ratio. These combined factors could be the reason for the peak power collected at $f_s/f_n = 1.6$ instead of 1.

Using the electromagnetic transducer, Jung et al. [42] designed and fabricated a harvesting device based on wake galloping. Based on the experimental findings of the wind tunnel, it has been shown that an average extracted power of 370 mW can be reached when the wind speed is 4.5 m/s, which offers a great advantage in the light wind conditions to power the bridge wireless sensor nodes.

11.4.1 Types of wake galloping

It is also known that major oscillations for wake galloping occur on a two-cylinder distance between $6D$ and $3D$, where D represents the cylinder diameter.

11.4.1.1 Wake galloping with unstable slope

There occurs a drastic increase when the distance of the spacing reaches $3D$, this is known as wake galloping with unstable slope.

11.4.1.2 Wake galloping with stable slope

However, the cylinder amplitude increases steadily with the gap between $4D$ and $6D$, and this type of wake-gallop is called wake-galloping with steady pitch.

It must be noted that without showing the computational/analytical comparisons, only experimental measurements were analyzed and presented.

Abdelkefi and his colleagues have experimentally studied the influence on the spectrum of flow rates of galloping energy harvesters. The authors then created a square-galloping-based energy harvester that produced energy at 0.3 to 0.4 m/s velocities and at low wind speed (< 1.7 m/s).

A wide spacing range between downstream and upstream with two different upstream circular cylinders was considered by Abdelkefi et al. [43]. The design of effective energy harvesters was proved to benefit from wake galloping. In particular, there can be a considerable increase in the speed range over which energy may be harvested. They considered that, depending on the harvester's placement (ventilation outlets, urban areas, lifting components in aircraft structures, river), with appropriate diameter selection, distance spacing between the circular cylinder and square section, and electrical load resistance, it is possible to design an enhanced piezoaeroelastic system.

11.4.2 Turbulence effects

Turbulence effects on the efficiency of a galloping energy harvester have been experimentally studied by Abdelkefi et al. [44]. The authors put

212   Piezoelectric Aeroelastic Energy Harvesting

meshes upstream of their harvester to study the effects of upstream turbulence on harvester reaction. The wake effects, along with upstream turbulence, have been reported to adversely influence the harvester's response. Their experimental findings showed that upstream and wake turbulence would improve harvest power positively, depending on the spacing between the two squares and the mesh opening size.

11.4.3 Galloping response

With a small non-linear term, Eq. (11.7) can be either solved by asymptotic methods or numerically. The damping and aerodynamic forces of order ζ and $\frac{U^*}{m^*}$, respectively, are small in comparison to the stiffness force (in the dimensionless equation of order unity) and inertia. The solutions of Eq. (11.7) tend to a limit cycle with normalized amplitude $A^* = \frac{A}{D}$ (oscillations' amplitude is denoted by A) quasi-harmonic oscillations. Air elastic bodies can benefit from this behavior where m^* is usually of the order of 10^3.

Eq. (11.7) is solved approximately by employing Krylov–Bogoliubov method, and one can find the function of the cross-section geometry (a_3 and a_1), the normalized amplitude of oscillations, mass, and mechanical properties ($m^*\zeta$ known as the damping frequency), and flow velocity:

$$A^* = (\frac{4U^*}{3a_3}(4m^*\zeta - a_1 U^*))^{\frac{1}{2}}. \tag{11.9}$$

From (11.9) it can be seen that $U^* > U_g^* = \frac{4m^*\zeta}{a_1}$ where $a_3 < 0$ and $a_1 > 0$. The linear version of Eq. (11.7) can retrieve U_g^*, and at U_g^* the effect of destabilizing fluid force equals the mechanical damping of the stabilizing effect. Hence it can be said that the galloping is possible only for some bluff bodies, with the stalled flow and for the low induced angle of attack values when the C_y slope is positive ($a_1 > 0$). For a body reattached to the lower shear layer with larger angle of attack values, and the C_y slope again becomes negative. It is noted that the low values of $m^*\zeta$ can make the galloping appear at low velocities.

11.5 Conversion factor

The ratio of power by flow to the body per unit length and total flow power per unit length is known as the conversion factor, or the efficiency, and is given by

$$\eta_1 = \frac{P_{F-B}}{P_F},$$ (11.10)

where the total power in the flow per unit length is

$$\frac{\rho U^3 D}{2}.$$ (11.11)

The extracted power by oscillating body from the flow per oscillation cycle T and per unit length is

$$P_{F-B} = \frac{1}{T} \int_0^T F_y \dot{y} dt.$$ (11.12)

Sinusoidal oscillations with amplitude A and frequency ω_N are given by

$$y = A \sin \omega_N t.$$ (11.13)

So, the conversion factor η_I in terms of reduced velocity and normalized amplitude can be expressed as

$$\eta_I = \frac{a_1}{2} \left(\frac{A^*}{U^*} \right)^2 + \frac{3a_3}{8} \left(\frac{A^*}{U^*} \right)^4.$$ (11.14)

On the right-hand side, the first term is positive where as the second term is negative ($a_1 > 0$, $a_3 < 0$). Inserting Eq. (11.9) into Eq. (11.14) gives

$$\eta_I = 2a_1 \left(\frac{4m^* \zeta a_1 U^*}{3a_3 U^*} + 6a_3 \left(\frac{4m^* \zeta a_1 U^*}{3a_3 U^*} \right) \right),$$ (11.15)

which is influenced by the cross-section and elastic properties, mass, and flow velocity in the conversion factor. Static aerodynamic characteristics of a square and isosceles cross-sections is represented in Table 11.1.

11.6 Evaluation of critical conditions with refined and multi-model approaches

Critical galloping conditions are typically studied for sectional versions, taking into account only the fundamental mode of vibration in each direction. In comparison, the effect of the mean wind force is still commonly ignored and neglected. Sometimes trivial conclusion throughout the evaluation of critical and post-critical principles is found (for example, in [48]). To this

Table 11.1 Static aerodynamic characteristics of a square and isosceles cross-sections.

Cross-section	$\eta_{I\,max} = \frac{-a_1^2}{6a_3}$	a_1	a_3	Reference and details
Square	0.05	2.3	−18	[45]; $33000 < Re < 66000$; $Tu \approx 0$
Isosceles triangle ($\delta = 30°$)	0.25	2.9	−6.2	[46]; $Re \approx 10^5$; $Tu = 4\%$
D-section	0.54	0.79	−0.19	[16]; $Re \approx 10^5$; $Tu = 11\%$
Isosceles triangle ($\delta = 53°$)	0.09	1.9	6.7	[47]; $Re \approx 10^4$

end, a step towards ensuring proper consideration of the deformed system setup is taken, as with monotube towers, for example, fitted with lighting or other rotating equipment, resulting in major effects on the flexible shaft. A wider structural model depending upon multi-body dynamics will also reinforce the classical part solution, i.e., dynamics of two or more rigid or fluid interconnected bodies [49] for a clearer explanation of the cable-stayed bridge cross-sections. More important concerns include the efficient impact of higher modes as the mechanical properties of the elastic system differ and the probability of seeking an estimated disruption solution for critical situations, considering a large number of Galerkin modes. A fascinating method of numerical-perturbation, depending upon eigenvalue sensitivity analysis, is built-in [50]; it is capable of constructing linear stability diagrams of multi-parameter dynamic systems representing a situation of potentially critical conditions.

A number of applications cannot be overlooked and should be restored for the torsion process. This is possible on two levels: aerodynamic forces' determination, provided the twisting does not explicitly affect the equilibrium conditions, or to derive a detailed model, including torsional velocity if the torsional DOF can substantially influence critical conditions or maybe unstable potentially. The latter feature poses a challenging and fascinating problem that is still frequently debated in the literature from different theoretical, experimental, and numerical points of view [51–53]. The authors of [54] proposed an instability model which is flow-induced, linear quasi-steady, 3-DOF, integrating inertial coupling between the DOF, and theoretically capable of calculating drag crisis instability and dry inclined cable galloping. As is common, this approach takes aerodynamic torsional damping at a radial distance that could be used to adapt the performance of the aerodynamic model to the observed instabilities.

11.7 Semi-analytical versus numerical solutions: non-linear galloping

Several studies on lighting poles have demonstrated a substantial improvement in mechanical damping with motion amplitude, which means that the structural motion can be self-bonding through dissipated energy when the system takes over a branch of deformation forcing it to bifurcate. It also shows a possible increase in galloping onset velocity. The presence of heavy machinery at the top, which could be subject to massive displacements, may contribute to second-order effects, necessitating the extension of aeroelastic contributions over linear terms.

Subjects usually dealing with a non-linear range, e.g., cable dynamic problems, are vulnerable to major developments. For example, interesting knowledge may be derived from a comparison of analytical and numerical approaches. In fact, the probability of improvement tends to be non-linear, galloping the sagged wires. As far as translational galloping is concerned, the number of active modes should be increased (e.g., from two to four) to provide skew-symmetric modes that can make a major contribution to the overall dynamics of the system [55]. Individual parameters, as well as the mean wind velocity, can be used to investigate phenomena of bifurcation from the standpoint of the principle of bifurcation. The ratio of sag-to-span, the initial angle of attack of the wind, and the structural modal damping all influence the ratio between the frequency and the critical velocity. A non-linear problem disruption analysis can be carried out from the optimum resonance state, in the chosen parameters of the discrete equation of motion, so that an explicit expression of the coefficient is expressed. The vicinity of the bifurcation point can be analyzed through parts of the space parameter, with a maximum possible dimension up to the co-dimension of the problem.

The creation of non-linear models of full 3D galloping is a real challenge not only with respect to aerodynamic non-linearity but also with the aid of consistent cable-like models, even geometric non-linearity [56]. These models can be overcome by using semi-analytical methods, such as the multi-scale perturbation system. A combination of numerical and analytical techniques has proven to be a very convenient method, as demonstrated in the preliminary analysis [57]. Several ideas can be found in the recent literature concerning particular finite element strategies that can be profitably used in the form of non-linear galloping oscillations [58–61]. Involved contributions can even emerge through a contrast of analytical/numerical outcomes and experimental evidence inferred from full-scale galloping [62] or wind tunnel research [63,64].

11.8 Harnessable energy

Energy harvesting is one of the critical tasks in micro- and nano-electromechanical systems [65–68]. Piezoelectric materials play a vital role in these energy harvesting systems [69–73]. These piezoelectric materials are widely used in the aerospace industry for energy harvesting, sensors, actuators, and structural health monitoring [74–77]. The difference of extracted power from the dissipated power in different stages of transmission can be converted into electricity. The case can be simplified by the assumption of internal losses, including both extracted energy by mechanical damping ζ and electric generator ζ_g (generator-associated damping). Harnessable energy P_{HE} per unit length is represented as

$$P_{HE} = \rho U^3 D [a_1 (\frac{4m^* \zeta_T a_1 U^*}{3 a_3 U^*}) + 3 a_3 (\frac{4m^* \zeta_T a_1 U^*}{3 a_3 U^*})^2] \frac{\zeta_g}{\zeta_T}, \qquad (11.16)$$

where $\zeta_T = \zeta + \zeta_g$ is the total damping. The overall energy harvested by a galloping-based aeroelastic energy harvester is represented in Fig. 11.5. It is observed that with the increase in R, the output average power P_{avg} increases along with increase in airflow velocity [9].

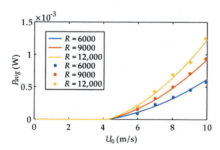

Figure 11.5 Output power of a galloping-based piezoelectric energy harvester [9].

References

[1] Hassan Elahi, Khushboo Munir, Marco Eugeni, Paolo Gaudenzi, Reliability risk analysis for the aeroelastic piezoelectric energy harvesters, Integrated Ferroelectrics 212 (1) (2020) 156–169.

Galloping-based aeroelastic energy harvesting   **217**

[2] Hassan Elahi, Marco Eugeni, Luca Lampani, Paolo Gaudenzi, Modeling and design of a piezoelectric nonlinear aeroelastic energy harvester, Integrated Ferroelectrics 211 (1) (2020) 132–151.

[3] Hassan Elahi, Marco Eugeni, Federico Fune, Luca Lampani, Franco Mastroddi, Giovanni Paolo Romano, Paolo Gaudenzi, Performance evaluation of a piezoelectric energy harvester based on flag-flutter, Micromachines 11 (10) (2020) 933.

[4] Hassan Elahi, The investigation on structural health monitoring of aerospace structures via piezoelectric aeroelastic energy harvesting, Microsystem Technologies (2020) 1–9.

[5] Marco Eugeni, Hassan Elahi, Federico Fune, Luca Lampani, Franco Mastroddi, Giovanni Paolo Romano, Paolo Gaudenzi, Numerical and experimental investigation of piezoelectric energy harvester based on flag-flutter, Aerospace Science and Technology 97 (2020) 105634.

[6] Hassan Elahi, Marco Eugeni, Paolo Gaudenzi, Design and performance evaluation of a piezoelectric aeroelastic energy harvester based on the limit cycle oscillation phenomenon, Acta Astronautica 157 (2019) 233–240.

[7] Marco Eugeni, Hassan Elahi, Federico Fune, Luca Lampani, Franco Mastroddi, Giovanni Paolo Romano, Paolo Gaudenzi, Experimental evaluation of piezoelectric energy harvester based on flag-flutter, in: Conference of the Italian Association of Theoretical and Applied Mechanics, Springer, 2019, pp. 807–816.

[8] Jacob Pieter Den Hartog, Mechanical Vibrations, Courier Corporation, 1985.

[9] Kaiyuan Zhao, Qichang Zhang, Wei Wang, Optimization of galloping piezoelectric energy harvester with v-shaped groove in low wind speed, Energies 12 (24) (2019) 4619.

[10] Galloping stability of triangular cross-sectional bodies: a systematic approach, Journal of Wind Engineering and Industrial Aerodynamics 95 (9) (2007) 928–940.

[11] Hysteresis in transverse galloping: the role of the inflection points, Journal of Fluids and Structures 25 (6) (2009) 1007–1020.

[12] Transverse galloping at low Reynolds numbers, Journal of Fluids and Structures 25 (7) (2009) 1236–1242.

[13] G.R. Benini, E.M. Belo, F.D. Marques, Numerical model for the simulation of fixed wings aeroelastic response, Journal of the Brazilian Society of Mechanical Sciences and Engineering 26 (2) (2004) 129–136.

[14] A. Laneville, et al., An explanation of some effects of turbulence on bluff bodies, 1977.

[15] Eduard Naudascher, Donald Rockwell, Flow-Induced Vibrations: An Engineering Guide, Courier Corporation, 2012.

[16] Milos Novak, Hiroshi Tanaka, Effect of turbulence on galloping instability, Journal of the Engineering Mechanics Division 100 (1) (1974) 27–47.

[17] G.V. Parkinson, Flow-induced structural vibrations, in: E. Naudascher (Ed.), 1974, pp. 81–127.

[18] Milos Novak, Aeroelastic galloping of prismatic bodies, Journal of the Engineering Mechanics Division 95 (1) (1969) 115–142.

[19] Antonio Barrero-Gil, G. Alonso, A. Sanz-Andres, Energy harvesting from transverse galloping, Journal of Sound and Vibration 329 (14) (2010) 2873–2883.

[20] Abdessattar Abdelkefi, Zhimiao Yan, Muhammad R. Hajj, Modeling and nonlinear analysis of piezoelectric energy harvesting from transverse galloping, Smart Materials and Structures 22 (2) (2013) 025016.

[21] A. Abdelkefi, Z. Yan, M.R. Hajj, Nonlinear dynamics of galloping-based piezoaeroelastic energy harvesters, The European Physical Journal Special Topics 222 (7) (2013) 1483–1501.

[22] Jayant Sirohi, Rohan Mahadik, Piezoelectric wind energy harvester for low-power sensors, Journal of Intelligent Material Systems and Structures 22 (18) (2011) 2215–2228.

218 Piezoelectric Aeroelastic Energy Harvesting

[23] Jayant Sirohi, Rohan Mahadik, Harvesting wind energy using a galloping piezoelectric beam, Journal of Vibration and Acoustics 134 (1) (2012).

[24] Abdessattar Abdelkefi, Muhammad R. Hajj, Ali H. Nayfeh, Power harvesting from transverse galloping of square cylinder, Nonlinear Dynamics 70 (2) (2012) 1355–1363.

[25] Yaowen Yang, Liya Zhao, Lihua Tang, Comparative study of tip cross-sections for efficient galloping energy harvesting, Applied Physics Letters 102 (6) (2013) 064105.

[26] Abdessattar Abdelkefi, Zhimiao Yan, Muhammad R. Hajj, Temperature impact on the performance of galloping-based piezoaeroelastic energy harvesters, Smart Materials and Structures 22 (5) (2013) 055026.

[27] Liya Zhao, Lihua Tang, Yaowen Yang, Comparison of modeling methods and parametric study for a piezoelectric wind energy harvester, Smart Materials and Structures 22 (12) (2013) 125003.

[28] Mohammed Ali, Mustafa Arafa, Mohamed Elaraby, Harvesting energy from galloping oscillations, in: Proceedings of the World Congress on Engineering, vol. 3, WCE, 2013, pp. 3–5.

[29] A. Bibo, M.F. Daqaq, On the optimal performance and universal design curves of galloping energy harvesters, Applied Physics Letters 104 (2) (2014) 023901.

[30] L. Tang, L. Zhao, Y. Yang, E. Lefeuvre, Equivalent circuit representation and analysis of galloping-based wind energy harvesting, IEEE/ASME Transactions on Mechatronics 20 (2) (2015) 834–844.

[31] Liya Zhao, Lihua Tang, Hao Wu, Yaowen Yang, Synchronized charge extraction for aeroelastic energy harvesting, in: Active and Passive Smart Structures and Integrated Systems 2014, vol. 9057, International Society for Optics and Photonics, 2014, 90570N.

[32] Liya Zhao, Lihua Tang, Yaowen Yang, Enhanced piezoelectric galloping energy harvesting using 2 degree-of-freedom cut-out cantilever with magnetic interaction, Japanese Journal of Applied Physics 53 (6) (2014) 060302.

[33] Zhimiao Yan, Abdessattar Abdelkefi, Muhammad R. Hajj, Piezoelectric energy harvesting from hybrid vibrations, Smart Materials and Structures 23 (2) (2014) 025026.

[34] Amin Bibo, Abdessattar Abdelkefi, Mohammed F. Daqaq, Modeling and characterization of a piezoelectric energy harvester under combined aerodynamic and base excitations, Journal of Vibration and Acoustics 137 (3) (2015).

[35] Zhimiao Yan, Abdessattar Abdelkefi, Nonlinear characterization of concurrent energy harvesting from galloping and base excitations, Nonlinear Dynamics 77 (4) (2014) 1171–1189.

[36] D. Vicente-Ludlam, A. Barrero-Gil, A. Velazquez, Optimal electromagnetic energy extraction from transverse galloping, Journal of Fluids and Structures 51 (2014) 281–291.

[37] H.L. Dai, A. Abdelkefi, U. Javed, L. Wang, Modeling and performance of electromagnetic energy harvesting from galloping oscillations, Smart Materials and Structures 24 (4) (2015) 045012.

[38] Masaru Matsumoto, Keisuke Mizuno, Kazumasa Okubo, Yasuaki Ito, Fundamental study on the efficiency of power generation system by use of the flutter instability, in: ASME Pressure Vessels and Piping Conference, vol. 47888, 2006, pp. 277–286.

[39] William Bradford Hobbs, Piezoelectric energy harvesting: vortex induced vibrations in plants, soap films, and arrays of cylinders, PhD thesis, Georgia Institute of Technology, 2010.

[40] William B. Hobbs, David L. Hu, Tree-inspired piezoelectric energy harvesting, Journal of Fluids and Structures 28 (2012) 103–114.

[41] Sylvain Dupont, Yves Brunet, Impact of forest edge shape on tree stability: a large-eddy simulation study, Forestry 81 (3) (2008) 299–315.

[42] Hyung-Jo Jung, Seung-Woo Lee, The experimental validation of a new energy harvesting system based on the wake galloping phenomenon, Smart Materials and Structures 20 (5) (2011) 055022.

[43] Abdessattar Abdelkefi, John Michael Scanlon, E. McDowell, Muhammad R. Hajj, Performance enhancement of piezoelectric energy harvesters from wake galloping, Applied Physics Letters 103 (3) (2013) 033903.

[44] Abdessattar Abdelkefi, Armanj Hasanyan, Jacob Montgomery, Duncan Hall, Muhammad R. Hajj, Incident flow effects on the performance of piezoelectric energy harvesters from galloping vibrations, Theoretical and Applied Mechanics Letters 4 (2) (2014) 022002.

[45] G.V. Parkinson, J.D. Smith, The square prism as an aeroelastic non-linear oscillator, The Quarterly Journal of Mechanics and Applied Mathematics 17 (2) (1964) 225–239.

[46] G. Alonso, J. Meseguer, I. Pérez-Grande, Galloping stability of triangular cross-sectional bodies: a systematic approach, Journal of Wind Engineering and Industrial Aerodynamics 95 (9–11) (2007) 928–940.

[47] S.C. Luo, Y.T. Chew, T.S. Lee, M.G. Yazdani, Stability to translational galloping vibration of cylinders at different mean angles of attack, Journal of Sound and Vibration 215 (5) (1998) 1183–1194.

[48] Robert D. Blevins, Flow-Induced Vibration, New York, 1977.

[49] Marco Lepidi, Giuseppe Piccardo, Aeroelastic stability of a symmetric multi-body section model, Meccanica 50 (3) (2015) 731–749.

[50] Angelo Luongo, Francesco d'Annibale, Linear stability analysis of multiparameter dynamical systems via a numerical-perturbation approach, AIAA Journal 49 (9) (2011) 2047–2056.

[51] W. Dettmer, D. Perić, A computational framework for fluid–rigid body interaction: finite element formulation and applications, Computer Methods in Applied Mechanics and Engineering 195 (13–16) (2006) 1633–1666.

[52] I. Robertson, L. Li, S.J. Sherwin, P.W. Bearman, A numerical study of rotational and transverse galloping rectangular bodies, Journal of Fluids and Structures 17 (5) (2003) 681–699.

[53] Thomas Andrianne, Experimental and numerical investigations of the aeroelastic stability of bluff structures, 2012.

[54] H. Gjelstrup, C.T. Georgakis, A quasi-steady 3 degree-of-freedom model for the determination of the onset of bluff body galloping instability, Journal of Fluids and Structures 27 (7) (2011) 1021–1034.

[55] Wen Yong Ma, Xiao Na Li, Ming Gu, Experimental investigation and theoretical analysis on galloping of iced conductors, in: Advanced Materials Research, vol. 860, Trans. Tech. Publ., 2014, pp. 1551–1558.

[56] P. Yu, Y.M. Desai, A.H. Shah, N. Popplewell, Three-degree-of-freedom model for galloping. Part I: Formulation, Journal of Engineering Mechanics 119 (12) (1993) 2404–2425.

[57] Angelo Luongo, Daniele Zulli, Giuseppe Piccardo, Analytical and numerical approaches to nonlinear galloping of internally resonant suspended cables, Journal of Sound and Vibration 315 (3) (2008) 375–393.

[58] Luca Martinelli, Federico Perotti, Numerical analysis of the non-linear dynamic behaviour of suspended cables under turbulent wind excitation, International Journal of Structural Stability and Dynamics 1 (02) (2001) 207–233.

[59] Y.M. Desai, P. Yu, N. Popplewell, A.H. Shah, Finite element modelling of transmission line galloping, Computers & Structures 57 (3) (1995) 407–420.

[60] Francesco Foti, Luca Martinelli, A corotational beam element to model suspended cables, in: 9th International Symposium on Cable Dynamics, 2011, pp. 1–8.

[61] Leopoldo Greco, Massimo Cuomo, An implicit G^1 multi patch B-spline interpolation for Kirchhoff–Love space rod, Computer Methods in Applied Mechanics and Engineering 269 (2014) 173–197.

[62] C.B. Gurung, H. Yamaguchi, T. Yukino, Identification of large amplitude wind-induced vibration of ice-accreted transmission lines based on field observed data, Engineering Structures 24 (2) (2002) 179–188.

[63] Olivier Chabart, Jean-Louis Lilien, Galloping of electrical lines in wind tunnel facilities, Journal of Wind Engineering and Industrial Aerodynamics 74 (1998) 967–976.

[64] Renaud Keutgen, J-L. Lilien, Benchmark cases for galloping with results obtained from wind tunnel facilities validation of a finite element model, IEEE Transactions on Power Delivery 15 (1) (2000) 367–374.

[65] Hassan Elahi, M. Rizwan Mughal, Marco Eugeni, Faisal Qayyum, Asif Israr, Ahsan Ali, Khushboo Munir, Jaan Praks, Paolo Gaudenzi, Characterization and implementation of a piezoelectric energy harvester configuration: analytical, numerical and experimental approach, Integrated Ferroelectrics 212 (1) (2020) 39–60.

[66] Hassan Elahi, Khushboo Munir, Marco Eugeni, Sofiane Atek, Paolo Gaudenzi, Energy harvesting towards self-powered IoT devices, Energies 13 (21) (2020) 5528.

[67] Hassan Elahi, Marco Eugeni, Paolo Gaudenzi, A review on mechanisms for piezoelectric-based energy harvesters, Energies 11 (7) (2018) 1850.

[68] Zubair Butt, Riffat Asim Pasha, Faisal Qayyum, Zeeshan Anjum, Nasir Ahmad, Hassan Elahi, Generation of electrical energy using lead zirconate titanate (PZT-5a) piezoelectric material: analytical, numerical and experimental verifications, Journal of Mechanical Science and Technology 30 (8) (2016) 3553–3558.

[69] Mohsin Ali Marwat, Weigang Ma, Pengyuan Fan, Hassan Elahi, Chanatip Samart, Bo Nan, Hua Tan, David Salamon, Baohua Ye, Haibo Zhang, Ultrahigh energy density and thermal stability in sandwich-structured nanocomposites with dopamine@Ag@BaTiO$_3$, Energy Storage Materials 31 (2020) 492–504.

[70] Ahsan Ali, Riffat Asim Pasha, Hassan Elahi, Muhammad Abdullah Sheeraz, Saima Bibi, Zain Ul Hassan, Marco Eugeni, Paolo Gaudenzi, Investigation of deformation in bimorph piezoelectric actuator: analytical, numerical and experimental approach, Integrated Ferroelectrics 201 (1) (2019) 94–109.

[71] Muhammad Usman Khan, Zubair Butt, Hassan Elahi, Waqas Asghar, Zulkarnain Abbas, Muhammad Shoaib, M. Anser Bashir, Deflection of coupled elasticity–electrostatic bimorph pvdf material: theoretical, fem and experimental verification, Microsystem Technologies 25 (8) (2019) 3235–3242.

[72] Vittorio Memmolo, Hassan Elahi, Marco Eugeni, Ernesto Monaco, Fabrizio Ricci, Michele Pasquali, Paolo Gaudenzi, Experimental and numerical investigation of PZT response in composite structures with variable degradation levels, Journal of Materials Engineering and Performance 28 (6) (2019) 3239–3246.

[73] Hassan Elahi, Marco Eugeni, Paolo Gaudenzi, Electromechanical degradation of piezoelectric patches, in: Analysis and Modelling of Advanced Structures and Smart Systems, Springer, 2018, pp. 35–44.

[74] Hassan Elahi, Khushboo Munir, Marco Eugeni, Muneeb Abrar, Asif Khan, Adeel Arshad, Paolo Gaudenzi, A review on applications of piezoelectric materials in aerospace industry, Integrated Ferroelectrics 211 (1) (2020) 25–44.

[75] Hassan Elahi, Marco Eugeni, Paolo Gaudenzi, Madiha Gul, Raees Fida Swati, Piezoelectric thermo electromechanical energy harvester for reconnaissance satellite structure, Microsystem Technologies 25 (2) (2019) 665–672.

[76] Hassan Elahi, Marco Eugeni, Paolo Gaudenzi, Faisal Qayyum, Raees Fida Swati, Hayat Muhammad Khan, Response of piezoelectric materials on thermomechanical shocking and electrical shocking for aerospace applications, Microsystem Technologies 24 (9) (2018) 3791–3798.

[77] Hassan Elahi, Zubair Butt, Marco Eugnei, Paolo Gaudenzi, Asif Israr, Effects of variable resistance on smart structures of cubic reconnaissance satellites in various thermal and frequency shocking conditions, Journal of Mechanical Science and Technology 31 (9) (2017) 4151–4157.

CHAPTER 12

Experimental aeroelastic energy harvesting

Contents

12.1. Introduction		223
	12.1.1 Ground vibrational tests	224
12.2. Experiments in a wind tunnel		225
	12.2.1 Testing of sub-critical flutter	226
	12.2.2 Flutter boundary	228
	12.2.3 Safety devices	231
	12.2.4 Research testing versus clearance testing	232
	12.2.5 Laws for scaling	232
	12.2.6 Flight experiments	233
	12.2.7 Flutter boundary approach	234
12.3. Testing of the flight flutter		234
	12.3.1 Excitation	234
12.4. Role of theory and experimentation in design		236
12.5. Experimental wing model		237
	12.5.1 Structural model	239
12.6. Slender body theory		240
12.7. Correlation between theory and experiment		241
12.8. Volterra theory experimentation		242
References		243

12.1 Introduction

Non-linear structures are of great concern to scientists from a diverse range of disciplines [1–8]. The research of non-linear processes begun in several ways in the pioneering work of Poincare, has developed steadily at the turn of the century, and is growing with the basic principles of non-linear mechanics being extended to various problems. The field of dynamical systems offers a coherent understanding of non-linear dynamics based on topological principles [9–11]. This mathematical explanation of non-linear dynamic processes helps scientists of diverse technical backgrounds.

Non-linear system identification approaches are used for addressing these issues in aeroelasticity, non-linear structural dynamics, and unsteady aerodynamics [12–16]. Lucia et al. [15] and Beran and Silva [12] provide information on new advances and implementations of the POD and HB

Piezoelectric Aeroelastic Energy Harvesting
https://doi.org/10.1016/B978-0-12-823968-1.00023-4

Copyright © 2022 Elsevier Inc.
All rights reserved.

methods. System identification techniques are classified into two types, parametric and non-parametric [17–20]. A parametric approach starts with a particular function model and specifies the coefficients that correspond to it. A non-parametric model is based on input/output maps such that a configuration is better represented functionally. Construction of reduced-order models (ROM) using CFD methodologies is an active area of research funded by several governments, industry, and academic institutions for computational aeroelasticity [16,21]. One of the ROM approaches currently under development is the development of ROMs based on the Volterra principle [22–25]. Piezoelectric materials are playing a vital role in these mechanisms [26–31]. They have wide applications in the field of aerospace industry [32–35] and for energy harvesting [36–39].

The underlying unsteady aerodynamics is best understood by experimental investigation of the dynamics of complex flight and phenomena of aeroelasticity. Experiments designed for the measurement of the response of unsteady aerodynamic response of different structures provide fascinating and useful information in this regard [40–43].

In-depth knowledge of the prevailing flow physics will contribute to the most effective flow control technique [44–46]. As a result, comprehending unsteady aerodynamic behavior is a critical phase in the development of flow control principles. This is the main objective of the research effort on flow management [47–50]. Higher-order spectra to flutter data [51] application and Volterra kernel identification from flight flutter experiments are extra examples of the Volterra theory [52–54].

Ueda et al. [55] advocate the use of specifying functions, a harmonic equilibrium technique of one harmonic, which is a modeling approach for unsteady transonic aerodynamic responses. Tobak et al. [56], by applying Volterra's continuous-time function description to indicial (step) aerodynamic responses, compute non-linear stability derivatives. Jenkins [57] also analyzes the determinations by using related functional principles of non-linear indicial aerodynamic responses and derivatives of non-linear stability.

12.1.1 Ground vibrational tests

Basic criteria of the ground vibration tests (GVT) include the excitation into the resonance of its structure, natural modes, and measurement of means of the response of the structure. A number of devices, including mechanical forces, electromagnetic forces, and acoustic excitation, have been used for excitation systems. Excitation choice mainly depends upon the required force level and frequency range that needs to be covered. For excited in the resonant mode lightly damped systems, estimation of the required force level can be done by the product of frequency with a square of critical damping and amplitude of the required response, i.e., stiffness and inertial terms almost cancel and balancing of a force of excitation is done by the damping. The required amplitude response is usually determined by instruments' sensitivity for measurement of available response or maybe the requirement of a certain linear response range for the measured response.

Mechanical excitation systems in practical terms are used for low frequencies from 1 to 100 Hz, acoustic excitation at high frequencies from 100 to 10,000 Hz, and shakers (electromagnetic exciters) for the moderate frequencies from 10 to 1,000 Hz.

Systems for the response measurement can be either electromagnetic (many electromagnetic devices can be used for either the response measurement devices or exciters), mechanical (accelerometers or strain gauges), or piezoelectric devices [58] which can be used for the responder or exciter.

In the basic technique of measurement, the system is excited at its resonant frequencies with devices of excitation and response, located at a structure position where a large response is expected. In order to differentiate between the natural mode that is anti-symmetric and symmetric, multiple exciters are used. In principle, excitation's continuous distribution with the force amplitude distribution proportional to the expected natural mode is optimum. To approach this ideal, many exciters can be used.

Using the pure frequency excitation, the modal's natural frequency is estimated, damping a transient decay [59]. Identification of the multiple modes with one excitation can be obtained by using a range of relevant frequencies with random excitation. For extraction of information, another possibility on multiple natural modes in the time domain is pulse excitation and using the fast Fourier transform. The latter measurements can be performed by using commercial software and hardware.

In the presence of significant non-linearities, to guide excitation form, data interpretation and measurement non-linear theory must be used. The number of opportunities is too large to be outlined conveniently.

12.2 Experiments in a wind tunnel

The environment needed for the excitation and response measurements in the presence of flow is complex, but the basic devices for the measurement, interpretation, and creation of the responses are the same as that of the experiments of simpler structural dynamics. Inside the tunnel, aerodynamic turbulence is present, which is often used for random excitation, there are no special requirements for the self-excited instabilities for exciting an aeroelastic system. However, it is preferable, where possible, to get an excitation device for conventional, e.g., acoustic, electromagnetic, piezoelectric, or mechanical, excitations. Experiments with sub-critical responses can be conducted using such excitations below the flutter boundary [60–62]. The schematics of the experimental campaign carried out in the wind tunnel are represented in Fig. 12.1 [63].

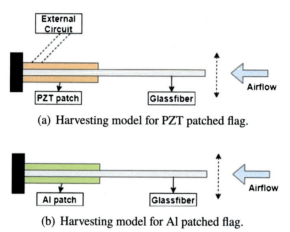

(a) Harvesting model for PZT patched flag.

(b) Harvesting model for Al patched flag.

Figure 12.1 Overall energy harvesting model of flag-flutter subjected to axial flow in sub-sonic wind tunnel [63].

12.2.1 Testing of sub-critical flutter

In the sub-critical flutter testing, modal damping change is monitored with the flow dynamic pressure change, for example, the dynamic pressure value anticipated for model damping becomes zero and negative. Moreover, for the rapid and dynamic pressure with complicated damping variation, the need for monitoring of critical modes makes it difficult extrapolating the flutter condition. Research on the extrapolation techniques is an active area of study for this reason [64].

For certain flutter types, the flutter onset can be predicted by monitoring modal frequency changes. Elahi et al. performed experimentation in the subsonic wind tunnel for energy harvesting via aeroelasticity as shown in Fig. 12.2. The dimensions of the flags in the experimental campaign are represented in Table 12.1 [63]. Flags with aluminum and piezoelectric patches are represented in Fig. 12.3 [63].

Table 12.1 Harvester's dimension in experimental campaign (cm) [63]; these flags are represented in Fig. 12.3 [63].

	Length	Width	Thickness
Al patched flag	15:3:38	6	0.05
PZT patched flag	29	6	0.05

(a) Experimental setup for Al patched flags.

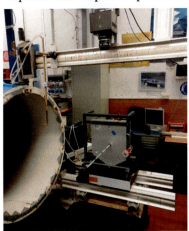

(b) Experimental setup for PZT patched flags.

Figure 12.2 Overall experimental campaign for flag-flutter energy harvesting mechanism carried out in sub-sonic wind tunnel [63].

Elahi et al. [63] performed an experimental campaign to analyze the aeroelastic energy harvesting based on LCOs. In this campaign, the fiberglass flag attached with piezoelectric patch (Fig. 9.1) and Al patch (Fig. 9.2)

228 Piezoelectric Aeroelastic Energy Harvesting

(a) Al patched prototypes.   (b) PZT and Al patched flags.

Figure 12.3 Al and PZT attached flags used for energy harvesting in sub-sonic wind tunnel [63].

are subjected to axial flow in a sub-sonic wind tunnel. When the airflow is increased up to a certain point, i.e., flutter velocity, the harvester starts flapping and moves away from LCOs as represented in Fig. 12.4 [63]. These experimental pictures are taken from FASTCAM and represents the flapping of aeroelastic harvester due to airflow. The output voltage generated by this experimental campaign is represented in Fig. 12.5 [63], it is observed that 0.68 MΩ external resistance is optimal for maximum energy harvesting. The amplitude of the voltage increases with the increase in external resistance up to certain resistance, i.e., optimal resistance, and after that the resistance will remain almost constant.

12.2.2 Flutter boundary

For incompressible flow (low speed) flutter sets, increasing the suitable increments of the flow velocity, the flutter boundary can be achieved. Compressible flow (high speed) flutter sets normally fix the Mach number, and, increasing stagnation pressure of wind tunnel, the flutter boundary is approached, and therefore in favorable increments the dynamic pressure, after which some changes are made to the Mach number followed by repeating

Experimental aeroelastic energy harvesting 229

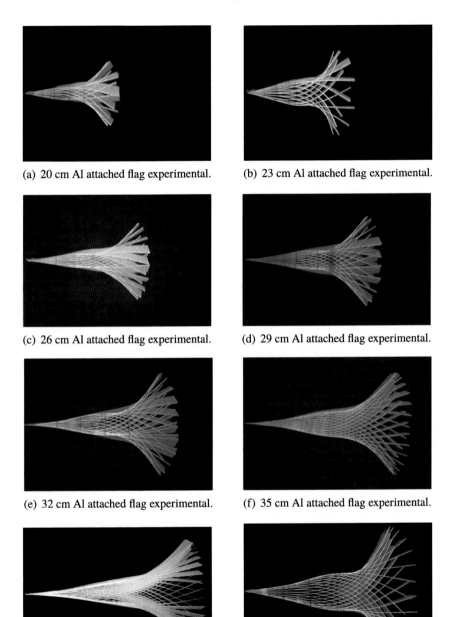

(a) 20 cm Al attached flag experimental.

(b) 23 cm Al attached flag experimental.

(c) 26 cm Al attached flag experimental.

(d) 29 cm Al attached flag experimental.

(e) 32 cm Al attached flag experimental.

(f) 35 cm Al attached flag experimental.

(g) 38 cm Al attached flag experimental.

(h) 29 cm PZT attached flag experimental.

Figure 12.4 Deformation comparison for Al and PZT patched flags from 15 to 38 cm long [63].

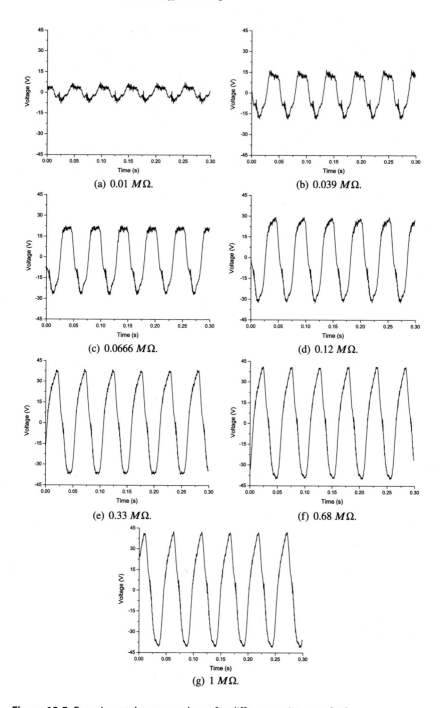

Figure 12.5 Experimental output voltage for different resistances [63].

the process. A blow-down wind tunnel at high Mach numbers seems to be the only available flow facility. It is preferred to have a tunnel with a closed return in a continuous flow when flow condition is well defined, and for the accurate response measurements an adequate time is obtained. The flutter boundaries for Al and PZT patched flags are analyzed by Elahi et al. [63], and they are represented in Figs. 12.6 and 12.7 for flutter velocity and flutter frequency, respectively.

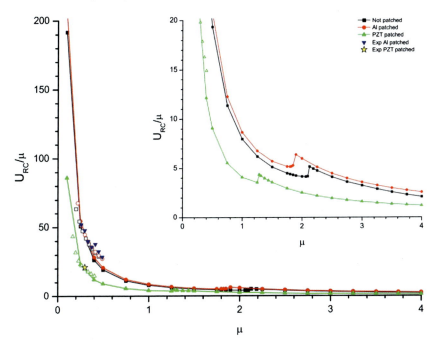

Figure 12.6 Flutter for reduced critical velocity [63], where $\mu = \rho_F L/\rho_S t$ and $U_{RC}/\mu = \left[(\rho_p h)^{3/2}/(\rho_F D^{1/2})\right] U$.

The experimental campaign for aeroelastic energy harvesting via galloping mechanism is presented in Fig. 12.8.

Moreover, the schematics for vortex-induced vibrations based energy harvesting is represented in Fig. 12.9.

12.2.3 Safety devices

Some provisions are needed for the flutter response suppression due to the rapid restraints to the model of flutter, for the protection of the model from flutter damage.

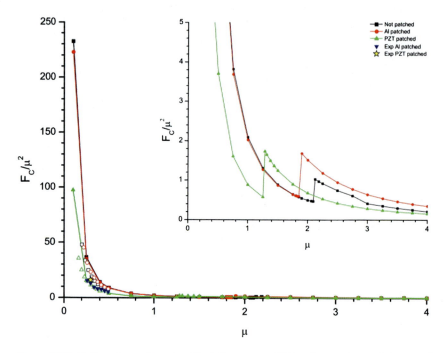

Figure 12.7 Flutter for reduced critical frequency [63], where $\mu = \rho_F L / \rho_S t$ and $F_C/\mu^2 = \left[(\rho ph)^{3/2}/(\rho_F D^{1/2})\right] f_c$.

12.2.4 Research testing versus clearance testing

For the comparison of the experimental data with the theory, some tests are conducted and detailed data sets are thus collected over a broad flow range and for various structural parameters. The reliability of the flutter model clearance tests is checked under a variety of predicted operating conditions.

12.2.5 Laws for scaling

The behavior of small scale models was studied inside the wind tunnel and can be linked by representing the aeroelastic motion equations in a dimensionless manner [67]. In view of the imitations of modal construction and wind tunnel flow conditions, it is not possible to match all the relevant dimensionless parameters between flight and tunnel scale. The choice of appropriate scaling parameters is an intelligent theory (i.e., matching the most sensitive and significant dimensionless parameters according to the analytical prediction) and experience-based judgment. The Mach number, reduced frequency, normally modal frequency ratios, and a dimensionless

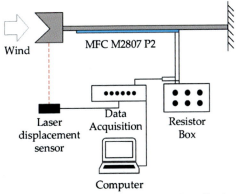

(a) Schematics of aeroelastic energy harvesting via galloping mechanism.

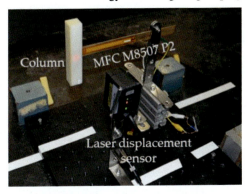

(b) Overall experimental campaign.

Figure 12.8 Experimental setup for aeroelastic energy harvesting via galloping mechanism [65].

ratio of dynamic pressure to model stiffness are matched, whereas structural mass/frequency fluid ratio is not.

Tests conducted in the wind tunnel are exceptionally valuable and fill in the knowledge gaps where theory is unreliable or unavailable.

12.2.6 Flight experiments

Almost all prior remarks about wind tunnel experiments refer to flight tests. The security requirement is crucial now, and the complexities of the provision of an excitation force which is well-defined are far greater. In addition, the procedure of the test is necessary. The flight flutter testing examples are represented in Table 12.2.

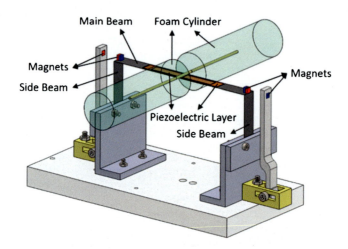

Figure 12.9 Experimental campaign for VIV based piezoelectric energy harvesting [66].

12.2.7 Flutter boundary approach

The flutter boundary was normally determined by altitude versus Mash numbers from the appropriate mixture of the analysis and the wind tunnel experiments before flight testing. Usually, with rising altitude, the Mach number responsible for flutter increases. The Mach number is raised in small increments at a fixed altitude before flutter occurs or the point where the aircraft reaches its maximum capacity of the Match number.

12.3 Testing of the flight flutter

Flutter testing is normally needed for modern aircraft in order to significantly alter current aircraft and introduce new applications of existing aircraft (Table 12.2).

12.3.1 Excitation

Several excitation techniques have been suggested and used. All are undeniably superior. Obviously, using existing hardware, such as electrical inputs to the control system, atmospheric turbulence, or control stick raps, reduces costs. Add-on devices, such as inertial mass oscillations, pyrotechnic devices, or oscillating vanes, on the other hand, presumably have greater control and excitation range. Reed [68] proposed a rotating slotted cylinder device which seems to be a reasonable balance between cost and efficiency.

Table 12.2 Flight flutter testing examples.

F-18 HARV AIRPLANE

1. Modification
Turning vanes–structures and control laws of the flight

2. Excitation
It commands the flight surface control

3. Testing
Testing the control systems of two different flights

4. Estimation
Over 80 test points are estimated for the clearance of the flight envelope

5. Angle of Attack
Attack angle range from $0°$ to $70°$

6. Mach Number
Mach numbers which are the multiples of points up to Mach 0.7

X-29 forward swept wing demonstrator

1. Testing
Control systems are tested with three different flights

2. Structure is new

3. Estimation
218+ test points are estimated for the clearance of the flight envelope

4. Excitation
Rotary inertia and turbulence shakers for the flaperons

AFTI/F-16 Aeroelastic and flutter test

1. Modification
Canards and digital flight control system

2. Excitation
Turbulence and stick raps

3. Stability Analysis
For the close spaced modes separation using the recursive identification algorithm

Schweizer 1-36 deep-stall sailplane

1. Modification
Modification of the horizontal stabilizer to pivot to $70°$

2. Excitation
Turbulence data at constant speed is acquired during the continuous descent in bands of altitude with test value of about $\frac{1}{2}1000$ feet

3. Testing
Before the flight tests, ground testing is performed due to the tail's non-linear structural dynamics

4. Stability Analysis
Monitoring of the strip charts cleared the stability analysis in real-time

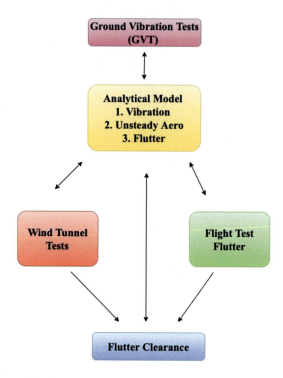

Figure 12.10 Aircraft's flutter clearance process.

Practical excitation systems have been used [69]. Aircraft's flutter clearance process is represented in Fig. 12.10.

12.4 Role of theory and experimentation in design

A combination of synergistic theory, wind tunnel testing, and flight tests is typically used in the design of a modern aircraft with appropriate aeroelastic behavior. A brief description of how this is often performed is given here. The cost of each of these synergistic components is one way to assess their relative value. For F-14 aircraft, Braid [69] estimated the values (29% analysis, 19% GVT, 27% wind tunnel, and 25% flight flutter test with total R&D cost of 0.5%).

It should be noted that each factor usually affects the others. The analysis and wind tunnel tests, for example, aid in the creation of the flight flutter test program.

Finally, although the focus has been on flutter studies, experiments of gust response or static aeroelastic activity can also be performed. The procedures used are similar to those for flutter tests, with the pilot and aircraft safety normally not being as important as for flutter tests.

12.5 Experimental wing model

The experimental model is composed of two components, namely a high aspect-to-ratio wing with a slender body on the top and a mechanism of root support. The wing is flexible, untwisted, and rectangular in the lag, flap, and torsional directions. The wing is precision ground with a mass that is uniformly spread over the wingspan in steel spar. The spar has width of 1.27 cm, length of 45.72 cm, and thickness of 0.127 cm. It is firmly incorporated into the mechanism of the wing root. The spar has some thin flanges along the span to decrease torsional rigidity. The flange is 0.318 cm deep and 0.127 cm in width. Along the centerline and wingspan of spar, flanges of 2 × 33 are symmetrically and uniformly distributed. Along the span, 18 pieces of NACA 0012 airfoil plate were uniformly distributed. These airfoil plate's pieces are composed of 0.245-cm thick aluminum alloy. The wing model's precise aerodynamic contour is achieved. At the symmetry line, the airfoil has a slot that is 0.127 cm thick and 1.27 cm in width. The spar is threaded into the slots in the airfoil plate and is firmly bonded to it. Each area between two airfoil plates is filled to provide the aerodynamical contours of the wing, with a light (bass) wood covering the entire chord. This wood has a slight extra weight and a modest increase in bending and torsion stiffness.

On the wing tip's elastic axis, a slender body is attached. A slender body is made up of aluminum bas, with 10.16 cm length and 0.95 cm diameter. An aft body and paraboloidal forebody with length 1.14 cm are fixed to the bar's two ends. The aft body and forebody are composed of brass. Paraboloidal forebody's geometry is described by

$$\frac{R}{R_0} = \bar{\gamma}^2, \qquad \bar{\gamma} = 0 \to 1. \tag{12.1}$$

The slender body is symmetric and has been established to have adequate torsional inertia to decrease the natural frequency of torsion which is essential for inducing the flutter in the wind tunnel's velocity range.

At the base, the angle of attack can be changed due to a socket which is basically the root support mechanism. On the sidewall of the wind tunnel, the root socket is mounted at the midpoint.

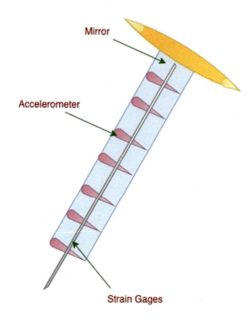

Figure 12.11 Physical representation of wing model.

Measurement of the wing's bending–torsional deflection is obtained by the root spar on which the strain gauge with 45-degree orientation for torsional modes was glued, along with the axial strain gauges for bending modes. Before taking measurements from the gauge conditioner and low pass filter, the signals coming from the strain gauges were conditioned and amplified. At the wing's midspan, a micro-accelerometer was mounted. A computer directly recorded the measurements from these transducers with data acquisition and analysis software, Lab-VIEW 5.1.

A random polarized 0.8 MW helium–neon laser with 633 nm wavelength is mounted on the tunnel's top. The tunnel's top is made of 1.27-cm thick glass plate. On the wing's tip, a mirror with a diameter 1.27 cm is fixed. The geometric twist angle is determined using the mirror deflection technique and, at the wing tip, the flaps or vertical bending slope. Physical representation of wing model is represented in Fig. 12.11. All LCO, flutter, and static tests were performed. The wind tunnel has a 0.7×0.53 m^2 test section and 1.52 m length and is a closed-circuit tunnel. Also the maximum attainable air speed is 89.3 m/s.

With the help of the standard static and vibration tests, the basic parameters of the experimental wing model were obtained.

12.5.1 Structural model

Neglecting the cross-sectional wing wrapping for untwisted elastic wing, the equations of motion can be written according to the Hodges–Dowell equations [70] as

$$EI_2 v'''' + (EI_2 - EI_1)[\phi(\omega)'']'' + m\ddot{v} + C_\zeta \dot{v} + (M\ddot{v} + I_v \ddot{v}')\delta(x - L)$$
$$= \frac{dF_v}{dx} + \Delta F_v - [Mg\delta(x - L) + mg]\sin\theta_0, \tag{12.2}$$

$$EI_1 v'''' + (EI_2 - EI_1)[\phi(v)'']'' + m\ddot{\omega} - me\ddot{\phi} + C_\zeta \dot{\omega} + (M\ddot{\omega} + I_\omega \ddot{\omega}')\delta(x - L)$$
$$= \frac{dF_\omega}{dx} + \Delta F_\omega - [Mg\delta(x - L) + mg]\cos\theta_0, \tag{12.3}$$

$$-Gj\phi'' + (EI_2 - EI_1)\omega'' v'' + I_\phi \ddot{\phi}\delta(x - L) + mK_m^2 \ddot{\phi} + C_\zeta \dot{\phi} - me\ddot{\omega}$$
$$= \frac{dM_x}{dx} + \Delta M_x. \tag{12.4}$$

If the above three equations are multiplied by δv, $\delta\omega$, and $\delta\phi$, respectively, followed by the integration over beam's length, then it is possible to derive a variational statement. These equations can also be derived from Hamilton's principle as in [70].

In the above three equations, it must be noted that Hodges–Dowell equations are used to retain the most important non-linear terms [70]. Geometrically, nonlinear third- and higher-order terms are neglected here. The angle of geometric twist in aerodynamic terms is considered to be

$$\hat{\phi} = \phi + \int_0^x v' \omega'' dx. \tag{12.5}$$

The tip mass is M, whereas the tip inertial terms are represented by I_v, I_ω and I_ϕ. The pitch angle or the steady angle of attack at wing model's root is given by θ_0, and the distance between wing elastic axis center and mass center is represented by e.

There are two contributions to aerodynamic forces. The first comes from the wing surface, $\frac{dF_v}{dx}$, $\frac{dF_\omega}{dx}$, and $\frac{dM_x}{dx}$, and the second is from the slender body which was attached at the wing tip, ΔF_v, ΔF_ω, and ΔM_x.

240 Piezoelectric Aeroelastic Energy Harvesting

12.6 Slender body theory

From the nomenclature given by Bisplinghoff et al. [71], $\alpha = \phi_{x=L}$, $h = \omega_{x=L}$, $M_x = \Delta M_x$, and $L = \Delta F_\omega$, whereas the chordwise position is denoted by y. Then at the wing tip the vertical displacement is defined as

$$Z_a = -h - \alpha[y - y_B], \tag{12.6}$$

where y_B is the distance of the slender body from the leading edge to the elastic axis. Positive up is defined by h and positive nose up is denoted by α. Now the convected vertical or downwash velocity is

$$W_a = \frac{\partial Z_a}{\partial t} + U\frac{\partial Z_a}{\partial y} = \frac{DZ_a}{Dt}, \tag{12.7}$$

where

$$D = \frac{\partial}{\partial t} + U(\frac{\partial}{\partial y}). \tag{12.8}$$

Thus,

$$\frac{dL}{dy} = -\rho\frac{DS}{Dt} - \rho S\frac{D^2 Z_a}{Dt^2} = -\rho\frac{D}{Dt}[S\frac{DZ_a}{Dt}], \tag{12.9}$$

where the body cross-sectional area is S and for a circular cross-section of radius $R(y)$ we have $S = \pi R^2$. Now

$$\frac{DS}{Dt} = U\frac{dS}{dy}, \tag{12.10}$$

so [71, Eq. (7-140)] becomes

$$\frac{dL}{dy} = -\rho U\frac{dS}{dy}\frac{Dz_a}{Dt} - \rho S\frac{D^2 Z_a}{Dt^2}. \tag{12.11}$$

Now,

$$L \equiv \int_0^{c_{SB}} \frac{dL}{dy} dy, \tag{12.12}$$

$$M_x \equiv \int_0^{c_{SB}} \frac{dL}{dy}[y - y_B] dy, \tag{12.13}$$

where the slender body is identical to the chord of length c_{SB}. Also

$$\frac{DZ_a}{Dt} = -\dot{h} - \dot{\alpha}[y - y_B] - U\alpha, \tag{12.14}$$

$$\frac{D^2 Z_a}{D^2 t} = -\ddot{h} - \ddot{\alpha}[y - y_B] - U\dot{\alpha}, \tag{12.15}$$

thus

$$\Delta F_\omega = L \equiv \int_0^{c_{SB}} \frac{dL}{dy} dy = \rho[\ddot{h} + U\dot{\alpha}] \int_0^{c_{SB}} S dy + \rho\ddot{\alpha} \int_0^{c_{SB}} S[y - y_B] dy, \tag{12.16}$$

$$\Delta M_x = M_x \equiv -\int_0^{c_{SB}} \frac{dL}{dy}[y - y_B] dy = \rho U[\dot{h} + U\alpha] \int_0^{c_{SB}} S dy$$

$$- \rho\ddot{h} \int_0^{c_{SB}} S[y - y_B] dy - \rho\ddot{\alpha} \int_0^{c_{SB}} S[y - y_B]^2 dy. \tag{12.17}$$

Note that

$$\int_0^{c_{SB}} \frac{dS}{dy} dy = -\int_0^{c_{SB}} S dy. \tag{12.18}$$

For a pointed body, $S = 0$ at $y = 0$.

Recall the convention of the signs, in which M_x is positive nose up, L-positive up, and y-positive aft.

Assuming that the drag of slender body can be neglected,

$$\Delta F_v = -(\phi_\lambda - \phi - \theta_0)\Delta F_\omega. \tag{12.19}$$

12.7 Correlation between theory and experiment

The experimental wing model's root mechanism is fixed on a very heavy support frame connected to the ground. The mechanism for root support requires the model to have the prescribed attack angle or pitch $\theta_0 = 0 \to 90$ degrees. Three special pitch angles are considered for assessing chordwise bending deflections, static flap, and twist at tip vs θ_0 under gravity loading, that is, $\theta_0 = 0, 45$, and 90 degrees. Measuring directly the difference between the deformed and undeformed tip position, the data is determined. By setting the aerodynamic term to zero in Eqs. (12.2), (12.3), and (12.4), the theoretical results can be obtained. To solve these equations and find a steady solution which is static deflection, a time-marching approach is used. The structural modes of the wing used in the study were four flap modes, one chordwise mode, and one torsional mode. Preliminary calculations using multiple modal configurations have confirmed that the analysis achieves strong convergence for these modes. The results obtained by Tang et al. [72, Figs. 3a–3c] showed tip flap and chordwise bending deflections and tip twist vs pitch angle. The authors compared the theoretical and experimental results by representing experimental results with solid dots and theoretical results with a solid line. The agreement shows outstanding agreement for the flap and chordwise deflection and is very strong for the twist.

The measurement of input source transfer function and acceleration or root strain of the output can determine the natural frequencies. Closer to the wing root, a B&K 8200 force transducer is fixed which is excited by B&K 4810 mini shaker and B&K 2706 power amplifier. A sweeping sinusoidal signal was provided by an SD 380 four-channel signal analyzer. The inputs to the SD 380 are the output signals from the chordwise micro-accelerometer and flap bending and torsion strain gauges for the analysis of transfer function. Using the experimental data from Dowell et al. [73], a correlation study was performed.

12.8 Volterra theory experimentation

The RSM has been used for measurement of the unsteady pressure, while at frequencies from 1 to 10 Hz, the model experienced pitch swings on the OTT. Furthermore, unsteady pressures were calculated during RSM/OTT step inputs to provide evidence for measuring aerodynamic impulse responses.

The same methods used to classify computational unsteady aerodynamic ROMs can be used to identify experimental unsteady aerodynamic (pressure) ROMs. The non-linear system Volterra theory, as defined in the references, serves as the basis for modeling the linear and non-linear dynamic response of the unsteady aerodynamic system.

The detection of experimental unsteady aerodynamic impulse responses will be restricted to the first-order, or linearized, kernel in this analysis. Since detection of the kernel (impulse response) can occur around a non-linear steady-state condition, it is referred to as a linearized kernel. The second-order kernel identification would be a subject of a future study.

The experimental unsteady aerodynamic impulse response is identified by using the given input/output deconvolution. In this case, a sequence of negative and positive steps is the input in pitch using the OTT, and any of the pressure responses measured from the wind-tunnel models can be the output. Then for the given input/output pair, the impulse response is extracted using deconvolution. For each pressure measurement, an impulse response can be identified for a given OTT step input.

At various frequencies, the pressure response due to sinusoidal inputs can be predicted using convolution after the generation of the impulse response [74]. Method validation is done by the comparison of the predicted and the measured results.

For the sake of brevity and to illustrate the method's viability, only one pressure measurement centered on the upper surface of the RSM at the position of 60% span and 30% chord station is provided. The data was collected at a dynamic pressure (q) of 150 psf, a Mach number (M) of 0.8, and an angle of attack of 0 degrees for the RSM. Theoretical step input is made of infinite slope where the steps occur with a physically realizable step input, commanded by OTT and limited by stress, pitch inertia, and limitation of a load of the model having pitch.

References

[1] Julius S. Bendat, Nonlinear Systems Techniques and Applications, Wiley, 1998.

[2] Hassan K. Khalil, Nonlinear Systems, Macmillan, New York, 1992.

[3] Ronald R. Mohler, Nonlinear Systems. Vol. 1: Dynamics and Control, Prentice-Hall, Inc., 1991.

[4] Francis C. Moon, Chaotic and Fractal Dynamics: Introduction for Applied Scientists and Engineers, John Wiley & Sons, 2008.

[5] A.H. Nayfeh, Perturbation Methods, Wiley, New York, 1973.

[6] Ali Hasan Nayfeh, Dean T. Mook, P. Holmes, Nonlinear oscillations, 1980.

[7] Rüdiger Seydel, Practical Bifurcation and Stability Analysis, vol. 5, Springer Science & Business Media, 2009.

[8] Lawrence N. Virgin, Introduction to Experimental Nonlinear Dynamics: A Case Study in Mechanical Vibration, Cambridge University Press, 2000.

[9] Jie Shen, Roger Temam, Nonlinear Galerkin method using Chebyshev and Legendre polynomials. I. The one-dimensional case, SIAM Journal on Numerical Analysis 32 (1) (1995) 215–234.

[10] Robert F. Brown, A Topological Introduction to Nonlinear Analysis, Springer, 1993.

[11] David K. Arrowsmith, Colin M. Place, C.H. Place, et al., An Introduction to Dynamical Systems, Cambridge University Press, 1990.

[12] Philip Beran, Walter Silva, Reduced-order modeling-new approaches for computational physics, in: 39th Aerospace Sciences Meeting and Exhibit, 2001, p. 853.

[13] Earl Dowell, John Edwards, Thomas Strganac, Nonlinear aeroelasticity, Journal of Aircraft 40 (5) (2003) 857–874.

[14] Epureanu Bogdan, Time-filtered limit cycle computation for aeroelastic systems, in: AIAA Atmospheric Flight Mechanics Conference and Exhibit, 2001, p. 4200.

[15] David J. Lucia, Philip S. Beran, Walter A. Silva, Reduced-order modeling: new approaches for computational physics, Progress in Aerospace Sciences 40 (1–2) (2004) 51–117.

[16] Walter A. Silva, Philip S. Beran, Carlos E.S. Cesnik, Randal E. Guendel, Andrew Kurdila, Richard J. Prazenica, Liviu Librescu, Piergiovanni Marzocca, Daniella E. Raveh, Reduced-order modeling: cooperative research and development at the NASA Langley Research Center, in: RECON, 2001, 20010071780, 16 pp.

[17] P. Eykhoff, System Identification-Parameter and State Estimation, John Wiley & Sons, London – New York – Sydney – Toronto, 1974.

[18] Jer-Nan Juang, Applied System Identification, Prentice-Hall, Inc., 1994.

[19] Ljung Lennart, System identification, in: Wiley Encyclopedia of Electrical and Electronics Engineering, 1999, pp. 1–19.

[20] Oliver Nelles, Nonlinear dynamic system identification, in: Nonlinear System Identification, Springer, 2001, pp. 547–577.

[21] Walter A. Silva, Moeljo S. Hong, Robert E. Bartels, David J. Piatak, Robert C. Scott, Identification of computational and experimental reduced-order models, in: International Forum on Aeroelasticity and Structural Dynamics Paper, 2003.

[22] Daniella Raveh, Yuval Levy, Moti Karpel, Aircraft aeroelastic analysis and design using CFD-based unsteady loads, in: 41st Structures, Structural Dynamics, and Materials Conference and Exhibit, 2000, p. 1325.

[23] Walter A. Silva, Application of nonlinear systems theory to transonic unsteady aerodynamic responses, Journal of Aircraft 30 (5) (1993) 660–668.

[24] Silva Walter, Reduced-order models based on linear and nonlinear aerodynamic impulse responses, in: 40th Structures, Structural Dynamics, and Materials Conference and Exhibit, 1999, p. 1262.

[25] Walter A. Silva, Discrete-time linear and nonlinear aerodynamic impulse responses for efficient CFD analyses, 1997.

[26] Hassan Elahi, Marco Eugeni, Paolo Gaudenzi, Electromechanical degradation of piezoelectric patches, in: Analysis and Modelling of Advanced Structures and Smart Systems, Springer, 2018, pp. 35–44.

[27] H. Elahi, A. Israr, R.F. Swati, H.M. Khan, A. Tamoor, Stability of piezoelectric material for suspension applications, in: 2017 Fifth International Conference on Aerospace Science & Engineering (ICASE), IEEE, 2017, pp. 1–5.

[28] Vittorio Memmolo, Hassan Elahi, Marco Eugeni, Ernesto Monaco, Fabrizio Ricci, Michele Pasquali, Paolo Gaudenzi, Experimental and numerical investigation of PZT response in composite structures with variable degradation levels, Journal of Materials Engineering and Performance 28 (6) (2019) 3239–3246.

[29] Muhammad Usman Khan, Zubair Butt, Hassan Elahi, Waqas Asghar, Zulkarnain Abbas, Muhammad Shoaib, M. Anser Bashir, Deflection of coupled elasticity–electrostatic bimorph PVDF material: theoretical, FEM and experimental verification, Microsystem Technologies 25 (8) (2019) 3235–3242.

[30] Ahsan Ali, Riffat Asim Pasha, Hassan Elahi, Muhammad Abdullah Sheeraz, Saima Bibi, Zain Ul Hassan, Marco Eugeni, Paolo Gaudenzi, Investigation of deformation in bimorph piezoelectric actuator: analytical, numerical and experimental approach, Integrated Ferroelectrics 201 (1) (2019) 94–109.

[31] Mohsin Ali Marwat, Weigang Ma, Pengyuan Fan, Hassan Elahi, Chanatip Samart, Bo Nan, Hua Tan, David Salamon, Baohua Ye, Haibo Zhang, Ultrahigh energy density and thermal stability in sandwich-structured nanocomposites with dopamine@Ag@BaTiO$_3$, Energy Storage Materials 31 (2020) 492–504.

[32] Hassan Elahi, Khushboo Munir, Marco Eugeni, Muneeb Abrar, Asif Khan, Adeel Arshad, Paolo Gaudenzi, A review on applications of piezoelectric materials in aerospace industry, Integrated Ferroelectrics 211 (1) (2020) 25–44.

[33] Hassan Elahi, Marco Eugeni, Paolo Gaudenzi, Madiha Gul, Raees Fida Swati, Piezoelectric thermo electromechanical energy harvester for reconnaissance satellite structure, Microsystem Technologies 25 (2) (2019) 665–672.

[34] Hassan Elahi, Marco Eugeni, Paolo Gaudenzi, Faisal Qayyum, Raees Fida Swati, Hayat Muhammad Khan, Response of piezoelectric materials on thermomechanical shocking and electrical shocking for aerospace applications, Microsystem Technologies 24 (9) (2018) 3791–3798.

[35] Hassan Elahi, Zubair Butt, Marco Eugnei, Paolo Gaudenzi, Asif Israr, Effects of variable resistance on smart structures of cubic reconnaissance satellites in various thermal

and frequency shocking conditions, Journal of Mechanical Science and Technology 31 (9) (2017) 4151–4157.

[36] Hassan Elahi, M. Rizwan Mughal, Marco Eugeni, Faisal Qayyum, Asif Israr, Ahsan Ali, Khushboo Munir, Jaan Praks, Paolo Gaudenzi, Characterization and implementation of a piezoelectric energy harvester configuration: analytical, numerical and experimental approach, Integrated Ferroelectrics 212 (1) (2020) 39–60.

[37] Hassan Elahi, Khushboo Munir, Marco Eugeni, Sofiane Atek, Paolo Gaudenzi, Energy harvesting towards self-powered IoT devices, Energies 13 (21) (2020) 5528.

[38] Hassan Elahi, Marco Eugeni, Paolo Gaudenzi, A review on mechanisms for piezoelectric-based energy harvesters, Energies 11 (7) (2018) 1850.

[39] Zubair Butt, Riffat Asim Pasha, Faisal Qayyum, Zeeshan Anjum, Nasir Ahmad, Hassan Elahi, Generation of electrical energy using lead zirconate titanate (PZT-5a) piezoelectric material: analytical, numerical and experimental verifications, Journal of Mechanical Science and Technology 30 (8) (2016) 3553–3558.

[40] Vladislav Klein, Patrick Murphy, Estimation of aircraft unsteady aerodynamic parameters from dynamic wind tunnel testing, in: AIAA Atmospheric Flight Mechanics Conference and Exhibit, 2001, p. 4016.

[41] D. Piatak, C. Cleckner, A new forced oscillation capability for the transonic dynamics tunnel, in: 40th AIAA Aerospace Sciences Meeting & Exhibit, 2002, p. 171.

[42] Robert Scott, Walter Silva, James Florance, Donald Keller, Measurement of unsteady pressure data on a large HSCT semispan wing and comparison with analysis, in: 43rd AIAA/ASME/ASCE/AHS/ASC Structures, Structural Dynamics, and Materials Conference, 2002, p. 1648.

[43] Walter Silva, Donald Keller, James Florance, Stanley Cole, Robert Scott, Experimental steady and unsteady aerodynamic and flutter results for HSCT semispan models, in: 41st Structures, Structural Dynamics, and Materials Conference and Exhibit, 2000, p. 1697.

[44] Marco Eugeni, Hassan Elahi, Federico Fune, Luca Lampani, Franco Mastroddi, Giovanni Paolo Romano, Paolo Gaudenzi, Experimental evaluation of piezoelectric energy harvester based on flag-flutter, in: Conference of the Italian Association of Theoretical and Applied Mechanics, Springer, 2019, pp. 807–816.

[45] Marco Eugeni, Hassan Elahi, Federico Fune, Luca Lampani, Franco Mastroddi, Giovanni Paolo Romano, Paolo Gaudenzi, Numerical and experimental investigation of piezoelectric energy harvester based on flag-flutter, Aerospace Science and Technology 97 (2020) 105634.

[46] Hassan Elahi, Khushboo Munir, Marco Eugeni, Paolo Gaudenzi, Reliability risk analysis for the aeroelastic piezoelectric energy harvesters, Integrated Ferroelectrics 212 (1) (2020) 156–169.

[47] M. Amitay, M. Horvath, M. Michaux, A. Glezer, Virtual aerodynamic shape modification at low angles of attack using synthetic jet actuators, in: 15th AIAA Computational Fluid Dynamics Conference, 2001, p. 2975.

[48] E. Chatlynne, N. Rumigny, M. Amitay, A. Glezer, Virtual aero-shaping of a Clark-Y airfoil using synthetic jet actuators, in: 39th Aerospace Sciences Meeting and Exhibit, 2001, p. 732.

[49] A.M. Honohan, M. Amitay, A. Glezer, Avia: adaptive virtual aerosurface, in: Fluids 2000 Conference and Exhibit, 2000.

[50] David Parekh, Ari Glezer, Avia – adaptive virtual aerosurface, in: Fluids 2000 Conference and Exhibit, 2000, p. 2474.

[51] Muhammad R. Hajj, Walter A. Silva, Nonlinear flutter aspects of the flexible HSCT semispan model, 2003.

[52] M.J. Brenner, R.J. Prazenica, Aeroservoelastic modeling and test data analysis of the F-18 active aeroelastic wing, in: International Forum on Aeroelasticity and Structural Dynamics, 2003.

[53] R. Lind, J.P. Mortagua, Extracting modal dynamics from Volterra kernels to reduce conservatism in the flutterometer, in: International Forum on Aeroelasticity and Structural Dynamics, 2003.

[54] R.J. Prazenica, M.J. Brenner, R. Lind, Nonlinear Volterra kernel identification for the F/A-18 active aeroelastic wing, in: International Forum on Aeroelasticity and Structural Dynamics, 2003.

[55] T. Ueda, E.H. Dowell, Flutter analysis using nonlinear aerodynamic forces, Journal of Aircraft 21 (2) (1984) 101–109.

[56] Murray Tobak, A study of nonlinear longitudinal dynamic stability, National Aeronautics and Space Administration (1964).

[57] Jerry Jenkins, Relationships among nonlinear aerodynamic indicial response models, oscillatory motion data, and stability derivatives, in: 16th Atmospheric Flight Mechanics Conference, 1989, p. 3351.

[58] Edward F. Crawley, Kenneth B. Lazarus, Induced strain actuation of isotropic and anisotropic plates, AIAA Journal 29 (6) (1991) 944–951.

[59] William Thomson, Theory of Vibration with Applications, CRC Press, 2018.

[60] Hassan Elahi, Marco Eugeni, Luca Lampani, Paolo Gaudenzi, Modeling and design of a piezoelectric nonlinear aeroelastic energy harvester, Integrated Ferroelectrics 211 (1) (2020) 132–151.

[61] Hassan Elahi, The investigation on structural health monitoring of aerospace structures via piezoelectric aeroelastic energy harvesting, Microsystem Technologies (2020) 1–9.

[62] Hassan Elahi, Marco Eugeni, Paolo Gaudenzi, Design and performance evaluation of a piezoelectric aeroelastic energy harvester based on the limit cycle oscillation phenomenon, Acta Astronautica 157 (2019) 233–240.

[63] Hassan Elahi, Marco Eugeni, Federico Fune, Luca Lampani, Franco Mastroddi, Giovanni Paolo Romano, Paolo Gaudenzi, Performance evaluation of a piezoelectric energy harvester based on flag-flutter, Micromachines 11 (10) (2020) 933.

[64] Yuji Matsuzaki, Yasukatsu Ando, Estimation of flutter boundary from random responses due to turbulence at subcritical speeds, Journal of Aircraft 18 (10) (1981) 862–868.

[65] Kaiyuan Zhao, Qichang Zhang, Wei Wang, Optimization of galloping piezoelectric energy harvester with V-shaped groove in low wind speed, Energies 12 (24) (2019) 4619.

[66] Wei-Jiun Su, Zong-Siang Wang, Development of a non-linear bi-directional vortex-induced piezoelectric energy harvester with magnetic interaction, Sensors 21 (7) (2021).

[67] John Dugundji, John M. Calligeros, Similarity laws for aerothermoelastic testing, Journal of the Aerospace Sciences 29 (8) (1962) 935–950.

[68] W.H. Reed III, Flight flutter testing: equipment and techniques, in: FAA Southwest Region Annual Designer Conference, Ft. Worth, Texas, 1991.

[69] Earl H. Dowell, Experimental aeroelasticity, in: A Modern Course in Aeroelasticity, Springer, 2015, pp. 479–486.

[70] Dewey H. Hodges, Earl H. Dowell, Nonlinear equations of motion for the elastic bending and torsion of twisted nonuniform rotor blades, 1974.

[71] R.L. Bisplinghoff, H. Ashley, R.L. Halfman, Aeroelasticity, 1955.

[72] Deman Tang, Earl H. Dowell, Experimental and theoretical study on aeroelastic response of high-aspect-ratio wings, AIAA Journal 39 (8) (2001) 1430–1441.

[73] E.H. Dowell, J. Traybar, Dewey H. Hodges, An experimental-theoretical correlation study of non-linear bending and torsion deformations of a cantilever beam, Journal of Sound and Vibration 50 (4) (1977) 533–544.

[74] Walter A. Silva, David J. Piatak, Robert C. Scott, Identification of experimental unsteady aerodynamic impulse responses, Journal of Aircraft 42 (6) (2005) 1548–1551.

CHAPTER 13

Concluding remarks

At the present time, the interest in data collection by means of large and distributed sensor networks is of increasing interest. Indeed, the modern technologies and Artificial Intelligence approaches permits a strong synergy between real/physical and software worlds, and this is demonstrated by the Industry 4.0 logics where Cyber-physical systems, Digital Twins and IoT representations of systems and processes are largely envisioned for monitoring and control purposes. It is clear that, in some situation/application is difficult or even impossible an external supplying of energy or the regular replacement on the depletion of the battery the possibility to self-power the sensors is a key factor in the future development of cited technologies. The employment of piezoelectric devices for energy harvesting purposes has attracted significant interest because it can be used to harvest energy over a wide range of frequencies and the ease of its application. Many researchers have used the piezoelectric transducer to develop simple and efficient energy harvesting devices from vibrations making the piezoelectric-based transducers promising in a wide range of applications. The possibility to extract energy by the interaction of a structure and the surrounding environment represented by an investing flow is one of the straightforward and easiest way of gathering energy from an operational environment. In this book the Authors presented a wide and deep description of the different architecture of energy harvesting approached based on aeroelastic phenomena showing which are the critical parameters that govern the design of such systems able to drive micro to nano electronics devices and demonstrating the importance that this kind of energy harvesting methods can play in the future technological scenarios.

Piezoelectric Aeroelastic Energy Harvesting
https://doi.org/10.1016/B978-0-12-823968-1.00024-6

Copyright © 2022 Elsevier Inc.
All rights reserved.

247

Subject index

0–9

2D aerodynamic surfaces, 132
2D beam, 131
2D modeling for static FSI system, 131
2D structures, 131

A

Abaqus, 99–101, 114, 117
Acoustic energy, 51
Aerodynamic
 kernel, 171
 lift, 160
 model, 139, 170
 modeling
  aeroelastic harvester, 169
 moment function, 136
 theories, 136
 transfer functions, 136
Aeroelastic
 damping, 169
 harvester, 168, 191, 228
  aerodynamic modeling, 169
  energy, 147
 model
  non-linear, 164
 problem, 128
 response, 164
 system, 125, 128, 158–160, 191, 226
  non-linear, 160, 162
Aeroelasticity, 125, 126, 136, 143, 144,
  223, 224, 226
 phenomenon, 130
Aeroelectroelastic state space equations, 175
Airflow, 147, 163, 170, 187, 228
Alternating current (AC), 51
Aluminum cantilevered shim, 191
Ambient energy, 79
Analysis
 flutter, 144, 175
  classical, 144
Angle of attack (AOA), 159
Antennas, 46

Assessment
 damage, 30
  laminated composites, 31
Assignment
 material, 104
 output, 110
Asymptotically stable system, 126

B

Base acceleration, 72, 195, 207
Beam
 2D, 131
 bending
  multiple, 68
  simple, 65
 cantilever, 13, 70, 112, 185–187, 206
 element, 65
 lift, 132
Bifurcation
 Hopf, 127, 128
 problems, 127
Bluff body, 185, 186, 190, 206, 209
 frequency, 189
 voltage, 206
Boundary
 conditions, 112
 flutter, 144, 226, 228, 231, 234

C

Calculation
 different from zero poles', 129
 power, 86
Cantilever beam, 13, 70, 112, 185–187,
  206
Cantilever-type vibration energy
  harvesting, 65
Circuit, 53, 81
 capacitive, 84
 corrective, 54
 designs, 92
 electronic, 11, 12
 energy
  conditioning, 82

249

250 Subject index

harvesting, 70, 81, 82, 86
equivalent, 85
electromechanical, 85
harvest, 84
harvesting, 71
piezoelectric energy, 81
impedance method, 87
MPPT, 88
rectification, 92
rectifier, 54
regulation, 71
resonant, 91
SSHI, 85, 91
synchronous, 82
topology, 83
Circuitry for enhanced energy harvesting, 84
Circular cylinder, 185, 190–192, 194, 211
Classical flutter analysis, 144
Classification, 31
Clearance testing, 232
Collocation point, 131
Condition
boundary, 112
critical galloping, 213
divergence, 129
loading, 112
Control
passive turbulence, 192
voltage, 82
system, 82
Converse effect, 7, 62
Conversion factor, 212
Corrective circuit, 54
Correlation, 241
Creation
input step, 108
job, 117
material, 103
Critical
galloping conditions, 213
mass, 184
Cubic polynomial, 204, 209
function, 203
terms, 204
Cylinder
circular, 185, 190–192, 194, 211

rigid, 194
electrode flexible ceramic, 191
rigid, elastically mounted spherical, 191
Cymbal type, 64

D

Damage assessment, 30
laminated composites, 31
Damping, 128, 174, 182–184, 194, 212, 216, 225
aeroelastic, 169
electromechanical, 194, 195, 207
frequency, 212
mechanical, 212, 215, 216
ratio, 191, 208
Damping/stiffness, 21
Degrees of freedom (DoF), 144
Density estimation, 31
Device
safety, 231
SSHI, 91
Direct numerical simulations (DNS), 193
Divergence, 127
condition, 129
Doublets lattice method, 130
Downwash, 133
Dynamic aeroelastic phenomenon, 126
Dynamic FSI, 133
Dynamical instability, 127

E

Eel concept, 189
Effect
converse, 7, 62
piezoelectric, 5, 8, 64, 70
turbulence, 211
Efficiency, 10, 48, 85, 192, 205, 211
energy, 14
harvesting, 51
harvester, 189
low boost converter, 92
maximum, 191, 194
power, 85
Electric
energy, 47, 48, 169
field, 7, 9, 23, 25, 62, 70
zero, 25

potential, 28, 29, 110, 117
power, 51, 64
voltage, 3, 5, 81
Electrical
   energy, 50, 53, 64, 67, 68, 79, 189
   load resistance, 194, 205, 207
   power output, 66
   resistance, 195
   voltage, 3, 64
Electrode flexible ceramic cylinder, 191
Electromagnetic induction, 46, 48, 192,
   206, 207
Electromechanical
   coupling coefficient, 73
   coupling constants, 62
   damping, 194, 195, 207
   energy transducers, 46
   equivalent circuits, 85
Electronic circuits, 11, 12
Electrorheological (ER) fluids, 23
Element
   beam, 65
   finite, 66
Energy
   acoustic, 51
   atmospheric, 79
   conditioning circuits, 82
   conversion
      efficiency, 66
      factor, 191
   coupling coefficient, 10
   demand, 41, 42, 51
   efficiency, 14
   electric, 47, 48, 169
   electrical, 50, 53, 64, 67, 68, 79, 189
   extraction, 201, 202
   generation, 13
   harnessable, 216
   harvest, 68, 190, 193, 195, 203
   harvester, 48, 66, 69, 190–192, 195, 206,
      207, 210
      flapping flag, 176
      piezoelectric aeroelastic, 152
   harvesting, 3, 9, 11, 42, 43, 46, 54, 61,
      65, 67, 68, 79, 100, 130, 151, 182,
      185, 189, 191, 192, 194, 202–204,
      216, 224

backpack, 68
cantilever-type vibration, 65
circuits, 70, 81, 82, 86
efficiency, 51
enhanced, 84
scheme, 42
techniques, 13
mechanical, 12, 13, 46, 48, 53, 67
PZT, 61
radio frequency, 53
recovery, 51
thermal, 49
transducer, 62
utilization, 15
vibration, 12, 46, 80
vortex-induced vibration aquatic clean,
   192
wind, 201
   harvesting, 170
Enhanced energy harvesting, 84
Equation
   aeroelectroelastic state space, 175
   constitutive, 8
   governing, 70
   motion, 144, 151
   non-linear electroelastic
      of motion, 173
Equivalent circuit method, 85
Estimation
   density, 31
   quasi-steady, 203
Excitation, 234
Experiment, 241
   flight, 233
   in a wind tunnel, 225
Experimental wing model, 237

F

Factor
   conversion, 212
      energy, 191
FEA, 12
Finite element (FE), 66
Finite element method (FEM), 3
   for flag-flutter, 130
Flag-flutter, 130
Flight experiments, 233

252 Subject index

Flight flutter, 224, 234
  test program, 236
  testing, 233
Flow velocity, 128, 145, 151, 212, 213, 228
Fluid–structure interaction, 48
Fluids
  electrorheological (ER), 23
  magnetorheological (MR), 23
Flutter, 127, 147
  analysis, 144, 175
  border, 144, 147
  boundary, 144, 226, 228, 231, 234
    approach, 234
  flight, 224, 234
  limit cycle, 159
  linear, 158, 159
  mechanism, 144
  onset, 226
  solutions, 146
  sub-critical, 226
  testing, 234
FOCV algorithm, 89
Frequency
  bluff body, 189
  damping, 212
  modal, 186, 226
  motion, 145
  ratio, 74
  vibration, 67, 184
  vortex shedding, 188, 189
FSI
  dynamic, 133
  system, 133
Function
  aerodynamic
    moment, 136
    transfer, 136
  Kussner, 138
  Wagner, 137, 140, 151

G

Galloping, 201, 202, 205, 206
  energy
    harvester, 205, 211
    harvesting, 204
  force, 204–207
    coefficients, 204

  non-linear, 215
  phenomenon, 202
  piezoelectric, 204
  response, 212
  transverse, 202, 208
  wake, 209–211
    phenomenon, 209
Generalized aerodynamic force (GAF), 131
Governing equations, 70
  piezothermoelastic response, 28
Ground vibration tests (GVT), 224

H

Harnessable energy, 216
Harvest
  circuit, 84
  energy, 68, 190, 193, 195, 203
Harvested
  power, 189, 195, 204, 206, 207, 210
    average, 73
    level, 207, 210
    output, 189
    value, 191
Harvester
  aeroelastic, 168, 191, 228
    energy, 147
  energy, 48, 66, 69, 190–192, 195, 206,
    207, 210
  flapping flag, 176
  galloping, 205, 211
  kinetic, 46
  piezoaeroelastic, 191
  piezoelectric aeroelastic, 152
  PZT, 64, 66
  sound, 53
  VIV, 189, 190
  performance enhancement, 192
  piezoelectric, 84, 85, 101, 108
    energy, 86, 92, 99, 100, 110, 115, 117,
    195
  PZT, 66
  response, 170, 207
  voltage, 82
    output, 82
Harvesting
  circuit, 71
    piezoelectric energy, 81
  electricity, 46

Subject index   253

energy, 3, 9, 11, 42, 43, 46, 54, 61, 65, 67, 68, 79, 100, 130, 151, 182, 185, 189, 191, 192, 194, 202–204, 216, 224
  cantilever-type vibration, 65
  electrical, 46
  enhanced, 84
  galloping, 204
  mechanical, 46, 47
  piezoceramics-based, 62
  piezoelectric, 11, 14, 42, 81, 85, 86, 131, 190
  PZT, 62, 65
  scheme, 42
  model, 70
  power, 61, 65, 69, 191
Hopf bifurcation, 127, 128
Hybrid energy scavenging system, 69

I

IFOCV algorithm, 89
Impedance matching, 84
Impedance matching network (IMN), 54
Impedance method circuit, 87
Inertia, 174
Input step creation, 108
Instability
  dynamical, 127
  static, 127
Interaction, 107
  fluid–structure, 48
Iterative method, 129

J

Job creation, 117

K

Kinetic, 44
Kinetic energy harvesters, 46
Kussner function, 138

L

Laws for scaling, 232
LCOs, 147, 157–159, 163, 227
  instabilities, 157
  response, 159
Lifting surface, 131, 139, 140, 145

Limit cycle
  flutter, 159
  oscillation, 157
Linear
  flutter, 158, 159
  response, 28, 225
Loading conditions, 112

M

Mach number, 126, 128, 135, 158, 159, 228, 231, 232, 234
Machine learning, 30, 31
  techniques, 31
Magnetorheological (MR) fluids, 23
Material
  assignment, 104
  creation, 103
  piezoelectric, 3, 11, 12, 15
  piezothermoelastic, 27
  PZT, 61–63, 65, 191
Maximum
  efficiency, 191, 194
  energy harvesting, 88, 89, 228
Maximum power point tracking (MPPT), 87
Mechanical
  damping, 212, 215, 216
    ratio, 191, 208
  energy, 12, 13, 46, 48, 53, 67
    harvesting, 46, 47
  modulation, 46, 48
    principle, 47
  stress, 3, 5, 6, 12, 65
  vibrations, 12, 13, 61, 65
Meshing, 113
Method
  doublets lattice, 130
  equivalent circuit, 85
  finite element, 130
  iterative, 129
  p–k, 128
Modal damping, 215, 226
Modal frequency, 186, 226
Model
  aerodynamic, 139, 170
  harvesting, 70
  non-linear aeroelastic, 164

254 Subject index

numerical, 130
structural, 172, 239
Motion
  equations, 144, 151
  frequency, 145
  pitch, 160–162, 167
  plunge, 148, 151, 161, 162
    structural stiffness, 167
MPPT, 87–89
  circuit, 88
  for piezoelectricity, 88
  performance, 89
  scheme, 88

N

Narrow frequency bandwidth, 85, 170
Natural frequency
  short circuit's, 73
Non-linear aeroelastic
  model, 164
  system, 160
Non-linear electroelastic equation of
  motion, 173
Non-linear galloping, 215
Numerical model, 130

O

On-linear aeroelastic
  system
    LCO, 162
Output assignment, 110

P

p–k method, 128
Passive turbulence control (PTC), 192
Phenomenon
  aeroelastic
    dynamic, 126
    static, 126
  aeroelasticity, 130
  galloping, 202
    wake, 209
Photovoltaic, 46
  technology, 50
Piezoaeroelastic energy harvester, 191
Piezoceramics, 24, 62
Piezoceramics-based energy harvesting, 62

Piezoelectric, 44
  cantilever, 85
    beam, 185, 201, 205, 209
    harvesters, 14
  constant, 70
  effect, 5, 8, 64, 70
  element, 29, 70, 71
  energy
    harvester, 86, 92, 99, 100, 110, 115,
      117, 195
    harvesting, 11, 14, 42, 81, 85, 86, 131,
      190
  galloping, 204
  generator, 85, 87–89, 91, 92, 170, 190
  harvester, 84, 85, 101, 108
  materials, 3, 11, 12, 15
  sensors, 24, 26
  smart structures, 22
  systems, 86
  transducer, 85, 169, 175, 202
  transduction, 11, 48, 100, 190
Piezoelectricity, 3–5, 7, 8, 61
Piezofilms, 24
Piezopolymers, 68
Piezothermoelastic materials, 27
Piezothermoelastic response governing
  equations, 28
Piston principle, 136
Pitch
  angles, 239, 241, 242
  motion, 160–162, 167
  variable, 161
Pitching moment, 145, 151, 167
Plunge, 160
  motion, 148, 151, 161, 162
  structural stiffness, 167
Power
  calculation
    for piezoelectric energy harvesting, 86
  efficiency, 85
  electric, 51, 64
  generator, 71
    array, 64
  harvested, 189, 195, 204, 206, 207, 210
    average, 73
    value, 191
  harvesting, 61, 65, 69, 191

output, 86, 204
transmission, 54
optimum, 87
wind, 46
Principle
mechanical modulation, 47
piston, 136
Problem
aeroelastic, 128
bifurcation, 127
PVDF, 24, 68, 69, 189
film, 33, 189
nanofibers, 69
PZT
energy, 61
harvester, 64, 66
harvesting, 62, 65
harvester, 66
materials, 61–63, 65, 191

Q

Quasi-steady, 139
estimation, 203

R

Radio frequency energy, 53
Rectification circuit, 92
Rectified voltage, 71, 72, 92, 93
Rectifier, 91
circuit, 54
Regression, 31
Regulation circuit, 71
Research testing, 232
Resonant circuit, 91
Response
aeroelastic, 164
characteristics, 159
galloping, 212
harvester, 170, 207
LCOs, 159
linear, 28, 225
Rigid circular cylinder, 194

S

Safety devices, 231
Sandwich structures, 26
SECE, 82, 84, 85, 91

Security threats, 54
Sensors
piezoelectric, 24, 26
wireless, 3, 15, 61, 79, 204
Shape memory alloys (SMA), 22
Shape/strain, 21
Short circuit's natural frequency, 73
Shunt diode, 85
Slender body, 140, 237, 240, 241
theory, 240
Smart structures, 23, 30
piezoelectric, 22
Sound energy
collection, 51
harvesters, 53
sources, 51
SSHI, 83–85, 91
circuits, 85, 91
device, 91
Stability, 126
Stable system, 127
Static
aeroelastic phenomenon, 126
FSI system, 131
instability, 127
Strip theory, 139
Structural model, 172, 239
Structures
2D, 131
sandwich, 26
Subcritical, 128
flutter, 226
Supercritical, 128
Surface
2D aerodynamic, 132
lifting, 131, 139, 140, 145
Synchronized switch harvesting on
inductor (SSHI), 91
Synchronous circuits, 82
System
aeroelastic, 125, 128, 158–160, 191, 226
energy
harvesting, 54
FSI, 133
non-linear aeroelastic, 160
LCO, 162
piezoelectric, 86

256   Subject index

stable, 127
  asymptotically, 126
static FSI, 131

T

Testing
  clearance, 232
  flutter
    flight, 233, 234
    sub-critical, 226
  research, 232
Tests
  ground vibrational, 224
Theory
  aerodynamic, 136
  slender body, 240
  strip, 139
  Volterra, 242
Thermal energy, 49
Thermoelectrics, 46
Transducer, 46, 48, 62, 88, 169
  energy, 62
    electromechanical, 46
  piezoelectric, 85, 169, 175, 202
Transverse galloping, 202, 208
Tunnel
  wind, 144, 210, 226, 228, 232, 233, 237, 238
    experiments, 225
    test, 159, 236
Turbulence effects, 211
Turbulent flow, 190

U

Unsteady
  aerodynamic, 224
    forces, 131, 185
    impulse responses, 242
    ROMs, 242
USECE, 85

V

Velocity–pressure relationship, 136
Vibration, 13, 14, 22, 27, 46, 68
  energy, 12, 46, 80
  frequency, 67, 184

mechanical, 12, 13, 61, 65
modes, 182
period, 72
vortex-induced, 185
VIV circular cylinder, 194
VIV energy harvesters, 189, 190
Voltage, 14, 22, 24, 89
  amplitude, 86
  bluff body, 206
  change, 24
  control, 82
    system, 82
  doublers, 93
  electric, 3, 5, 81
  electrical, 3, 64
  harvester, 82
  multiplier circuit, 54
  output, 82, 170, 195
  rectified, 71, 72, 92, 93
Volterra theory experimentation, 242
Vortex shedding frequency, 188, 189
Vortex-induced vibration aquatic clean
    energy (VIVACE), 192
Vortex-induced vibrations, 185

W

Wagner function, 137, 140, 151
Wake galloping, 209–211
  phenomenon, 209
Wavelet packet transform (WPT), 34
Wind, 46
  energy, 201
    harvesting, 170
  power, 46
  tunnel, 144, 210, 226, 228, 232, 233, 237, 238
    experiments, 225, 233
    test, 159, 236
Wing, 140, 160, 163, 164, 237
  behavior, 148
  model
    experimental, 237
  surface, 239
Wireless power transfer (WPT), 53
Wireless sensors, 3, 15, 61, 79, 204

Printed in the United States
by Baker & Taylor Publisher Services